T0283397

BUILDING A GOD

The Ethics of Artificial Intelligence and the Race to Control It

CHRISTOPHER DICARLO, PH.D.

 Prometheus Books

Essex, Connecticut

 Prometheus Books

An imprint of The Globe Pequot Publishing Group, Inc.
64 South Main Street
Essex, CT 06246
www.globepequot.com

Distributed by NATIONAL BOOK NETWORK

British Library Cataloguing in Publication Information Available

Library of Congress Cataloging-in-Publication Data

Names: DiCarlo, Christopher, 1962– author.
Title: Building a god : the ethics of artificial intelligence and the race to control it /
   Christopher DiCarlo.
Description: Lanham, MD : Prometheus, [2025] | Includes bibliographical references
   and index. | Summary: "In Building a God, Christopher DiCarlo, a global leader in
   the ethics of artificial intelligence, unpacks the tangled web surrounding AI, revealing
   to readers what we know, what we don't, and how we might prepare ourselves for
   eventualities that we don't know we don't know yet"—Provided by publisher.
Identifiers: LCCN 2024018294 (print) | LCCN 2024018295 (ebook) | ISBN
   9781493085880 (cloth) | ISBN 9781493085897 (epub)
Subjects: LCSH: Artificial intelligence—Moral and ethical aspects.
Classification: LCC Q334.7 .D53 2025 (print) | LCC Q334.7 (ebook) | DDC
   174/.90063—dc23/eng20240802
LC record available at https://lccn.loc.gov/2024018294
LC ebook record available at https://lccn.loc.gov/2024018295

This book is dedicated to my sons Jeremy and Matthew.
I cannot begin to thank you enough for the life you have given me.

Contents

Introduction

When building a god, we must be very, very, careful, for we are at the crossroads of perhaps our most unique moment in our short history as a species. For the first time in our existence, we possess the curiosity, capacity, and greed to create an artificially intelligent being (or beings) so intelligent and so powerful there is a probable likelihood that its construction may bring about the end of our own existence.

The major gods of past and current cultures have always been a top-down form of metaphysical being that possesses qualities and characteristics of being all-intelligent (omniscience), all-powerful (omnipotence), ever-present (omnipresence), and all-good (omnibenevolence). These gods were and are believed to have existed prior to ours and were considered to be not only creative forces but also prescribers of moral precepts to govern the actions of our species. With the increasingly rapid advancements in artificial intelligence (AI), we are witnessing a bottom-up approach wherein it is us, the humans, who have to figure out what moral precepts and commands should go into the making of such a being so that it will not turn against us and harm us in various ways—intentionally or accidentally. To avoid this Frankenstein Effect, building a god in our very best image is going to be our greatest collective challenge.

Given enough computing power, data, and time, it is inevitable that at some point in our not-too-distant future we will succeed in creating a form of intelligence that far surpasses our own. And when that time comes—and it really appears to be a matter of when, not if—how will such a being respond to us? How will such a god-like being react to our attempts for control and guidance? The answer is a bit unsettling: No one knows.

Figure I.1.   White to grey to black: "Known knowns, known unknowns, and unknown unknowns." *Created for the author. All rights to the author.*

What exactly do we know? Well, this describes what we call our current epistemic stance. In other words, what state of knowledge, or lack thereof, are we currently in? What do we know about the potential risks and harms of AI, and what don't we know? To demonstrate this I'll quote, of all people, Donald Rumsfeld—George W. Bush's secretary of defense. Rumsfeld once said: "There are known knowns—there are things we know we know. We also know there are known unknowns—that is to say, we know there are some things we do not know. But there are also unknown unknowns, the ones we don't know we don't know" (see figure I.1).[1]

Throughout this book, we will consider, very carefully, our epistemic stance by reflecting on what we know about AI and what we know we don't know about it, and then try to prepare ourselves for eventualities that we don't even know we don't know about what may come.

Toward this end, we are hopeful that our new, emerging, superintelligent machine god (or SMG) may fix some of our most pressing problems, like solving world hunger, curing diseases, stopping and reversing climate change, and, in general, making human and animal lives better. This is what everyone in the AI business wants. Such a god-like machine would deliver humanity from past inequities and mundane drudgeries and promote well-being throughout the world. Just think of it: a future where much of our manual-labored jobs are taken care of, where health care is affordable and available throughout the world, where energy is cheap and efficient. The developments of AI promise a future filled with considerable possibilities for the betterment of humankind, the environment, and so many other species. But there is absolutely no guarantee that an artificially intelligent god-like being will always be good—or that it will always be used by humans for the better good.

The DC comic hero Superman, for example, did not have to be a law-abiding, tax-paying citizen. He could have "smashed through any bank in the United States" yet refrained from doing so (as well as a lot of other horrendous acts against humanity). But who's to say a highly evolved and superintelligent form of AI would behave according to our ethical standards? Why would it? And what reasons do we have to believe it would? This leads us to consider the possibility of a superintelligent AI evolving as an existential threat or risk to us (known as x-risk). Will this occur? We don't know. But it's probably a safe bet to err on the side of caution since we are playing with technologies whose effects for the most part are unknown to us. So, if we were to imagine what this would look like in the form of gambling on our future, our best wager would look something like what is shown in figure I.2.

If we believe the development of a form of superintelligent AI may pose an existential risk to humanity at some point in the future, and it turns out to be true, then we as a species can win, so to speak, by preparing for this potentiality and guarding against its anticipated harmful effects. And if it turns out that, no matter how intelligent our AI technologies become in the future, they never pose an x-risk to us, then we would have taken the most epistemically and ethically responsible precautions to ensure against this. On the other hand, if we don't take x-risk seriously and there turns out to be no risk whatsoever, we would be acting irresponsibly—both epistemically and ethically—because we simply got lucky. And the worst case scenario occurs if we don't take x-risk seriously and it turns out to be

	X-RISK	NO AI X-RISK
YOU BELIEVE AI X-RISK IS POSSIBLE (Take action)	WIN	Neutral: Epistemically and Ethically Responsible
YOU DON'T BELIEVE AI X-RISK IS POSSIBLE (Do nothing)	LOSE	Neutral: Epistemically and Ethically Irresponsible

Figure I.2.   AI x-risk wager. *Created by the author.*

true; then we will face considerable harms and repercussions at some point in the future.

As you can imagine, there are those on either end of the spectrum who believe in the absolute safety of AI advancements—we'll call these folks the naysayers. And there are those who fall onto the opposite end of the spectrum who believe that our own self-destruction at the hands of superintelligent AI is a done deal—we'll call these people doomsayers. And then, of course, there are plenty who fall somewhere in between. The range of the degree of danger can be summed up by noting how everyone lies somewhere between Y2K and Armageddon. By the end of this book, you should have a better understanding of where, exactly, you fall on this spectrum, and why.

But what if there's only, say, a 5 percent chance that we humans will create such a powerful artificial superintelligence that will harm us or wipe out humanity entirely? Well, would you get on an airplane if you always faced a 5 percent chance of fatally crashing? Think about it: there's a 1 in 20 chance that every time you fly, your plane will crash and you will die. How often would you be getting on an airplane? Probably not often; if at all. So it's important to realize that we really want to make sure there's nowhere near a 5 percent chance of such an occurrence.

And it is exactly because we don't know, precisely, what the odds are in building a destructive form of AI that we must be very, very careful when considering how to build such a god. Because if we get it wrong, and there turns out to be a likelihood of potential existential risk, we may not have the time to control, contain, or stop its negative effects. So how do we do this? This leads us to ask ourselves *the* most central, pressing question facing all scientists, academics, politicians, and industry leaders in the AI business today: *How do we build a superintelligent machine—a mechanical god—that only helps us and can never harm us?*

If we had the answer, the world would be in a much more secure place, and a lot more people in the AI safety and governance business would be sleeping a whole lot better. But unfortunately, neither I, nor anyone on this planet, know how to do this. At least, not at the moment. But we are working on it. The team I lead—as well as our other teams at Convergence Analysis (a UK-based organization devoted to mitigating the existential risk of AI)—is working on better understanding current developments in AI technologies so we can advise industry leaders, politicians, and the public about the safest path to take moving forward. As our CEO, David Kristofferson believes:

The future of humanity could be flourishing or catastrophic: the outcome is up to what we do now. What is the nature of this problem we're trying to solve? Decision making. Human intelligence drives human civilization, and it drives it through decisions. Humans make decisions, and we act following decisions. Decisions determine what kind of AI we build and what we use it for. What kind of AI we build and what we use it for determines whether AI will be good or bad. Additionally, the act of building AI is the act of asking the machine to make decisions. And we'd better know well what we are asking it to do. In sum: We need humans to make the right decisions in developing and using AI, and we need the AI that we build to use the right decision-making.[2]

At this very unique point in our history, it is crucial for us to make the right decisions about why and how to build a god. Because if we get it wrong, it will be the worst and last machine we ever build.

So there is a great sense of urgency to spread the word very far, very wide, and very quickly that there is a race on with Big Tech and world political powers to be the first to build a superintelligent machine god (SMG). So what does that say about the future of humanity? It means that we need to focus more attention on this problem than we have on any other collective problem facing our existence. The issue of controlling the effects of AI is greater than putting humans on the moon or into space; greater than battling climate change; greater than world hunger; and even greater than nuclear annihilation. And if we don't act soon and collectively, the problem will be upon us quicker than we anticipated—leaving us little to no time to react.

So let me be very clear: the threat from AI of existential risk and annihilation to humanity is very real. The odds that it will occur are difficult to determine because, as mentioned earlier, no one knows for sure what is going to happen or precisely how soon. But we do know that we all want the best that AI has to offer us while eliminating any potential existential risks or harms. Our best shot at assuring this depends on how well we can critically think about such a threat and to establish universal, ethical guidelines that ensure the future safety of humankind. Throughout the remainder of this book, we will consider whether or not this is possible.

AI AND WHAT WE
SHOULD KNOW ABOUT IT

I

AI 101   1
Concepts, History, and Applications of AI

In this chapter, we will consider the basics of artificial intelligence (AI). We will explore the foundational aspects of AI to ensure that everyone shares a common understanding of important terms, concepts, and ideas. The field of AI is rapidly evolving and changing day by day, so it's crucial for us to grasp the key historical ideas and developments that have brought us to our current state of concern. This contextual background knowledge will empower us to have more informed and constructive discussions regarding the future path of AI with respect to its potential benefits, risks, and governance.

To understand the basics of AI, we will first examine its definition. What do we mean when we're talking about intelligence being "artificial"? How does it differ from, say, human intelligence? Once we have a better understanding of some of the key concepts behind the definition of AI, we can then look at its history—where did it come from? When did it start? Where will it lead us? Answering questions such as these will grant greater insight and capacity when considering deeper issues involving the applications of emerging AI technologies.

What Is Artificial Intelligence?

When you hear the words "artificial" and "intelligence," what comes to mind? What are you thinking about? Do you picture the unblinking red eye of the HAL 9000 computer in *2001: A Space Odyssey*? Do you hear the voice of Arnold Schwarzenegger in *The Terminator*, or the tragic final soliloquy delivered by Rutger Hauer, a Nexus 6 replicant, in *Blade Runner*? Or what about some of the more intense robotic creations found in the movie *Ex Machina*? Or are you thinking of Siri? ChatGPT? Alexa? Or your Roomba? Whatever your thoughts are about these two words, you probably have some idea of what is meant when they are used together. Let's look at them a bit more closely.

What does it mean for intelligence to be "artificial"?

Most dictionary definitions define the term "artificial" as a copy of something "natural" that is "made by humans." Some definitions even go so far as to call it "fake" or "not real."[1] For our purposes, and within the AI field itself, the term is used to distinguish it from human or animal forms of intelligence. And yes, this artificial form of intelligence is indeed created by humans; but the processes through which it is created are not similar to the various processes through which human or other forms of animal intelligence evolved. This distinction—that humans have endured a long and arduous evolutionary process whereas all forms of AI have been created in an instant—is an important one and should be kept in mind as we proceed through the remainder of this book.

So then, what does "intelligence" mean?

That is a much trickier question to answer. There are literally dozens of ways such a term can be defined.[2] We could simply respond with a basic dictionary definition stating that it involves the ability to learn, understand, and make judgments or have opinions that are based on reason,[3] but we know that it involves much more than this. It also involves the ability to learn or understand or deal with new or trying situations; it involves using not only reason but also creativity, as well as the ability to apply knowledge to navigate through and manipulate one's environment. At other times, it can also mean to think critically and abstractly as measured by objective criteria such as tests. And in some cases, it refers to the capacity for mental acuteness or shrewdness.[4] On the artificiality of intelligence, Pamela McCorduck writes:

> Intelligence means literally "to choose among what has been gathered." This etymology casts the history of artificial intelligence into a vastly larger landscape, revealing it to be not some hubristic overreaching (though it has sometimes seemed so), but instead another natural stage in the flow of that most enduring, even noble, of human urges: the passion to gather, organize, and share knowledge so that we all benefit. If competition typifies human behavior (and it does) then alongside it, possibly in equal measure, is cooperation. From the beginning, it seems, we've gathered and organized information, knowledge, and techniques largely to share all these with our fellow humans, to amplify their knowledge as well as our own.[5]

I believe McCorduck hits the proverbial nail right on the head when she talks about the equal measures for both competition and cooperation in gathering and sharing information. There are many studies and theories about the nature of intelligence, but for our purposes, when we combine elements of these various definitions together, we can define "artificial intelligence" as

> a nonnaturally electronic analogue of intelligence that can learn, understand, and make judgments based on reason, shrewdness, and creativity, which allows it to navigate through and manipulate a given environment critically and abstractly.

Although this provides some clarity, we still need to add a few things. The term has been defined elsewhere by notable figures such as AI pioneer and Stanford professor John McCarthy, who defined artificial intelligence as

> the science and engineering of making intelligent machines, especially intelligent computer programs. It is related to the similar task of using computers to understand human intelligence, but AI does not have to confine itself to methods that are biologically observable.[6]

And decades before McCarthy's definition, Alan Turing, the British father of computer science, was considering whether or not machines could think in his groundbreaking 1950 work "Computing Machinery and Intelligence." He devised the now famous "Turing Test" in which a human interrogator attempts to distinguish between a computer and human text response. If the interrogator is unable to distinguish between them, the computer is said to have passed the "intelligence" test.[7]

In the late 1990s, Stuart Russell and Peter Norvig, two prominent AI scientists, authored an innovative work titled *Artificial Intelligence: A Modern Approach*, which has since established itself as the preeminent textbook in the field of AI studies. In this seminal work, they outline four potential objectives or definitions of AI that distinguish computer systems based on their capacity for rationality and decision-making versus their capacity for action.

To start, they explore the "human approach," wherein the performance of systems that emulate both human thinking and human actions is assessed. This is juxtaposed against an ideal approach in which systems are expected to demonstrate rational thinking and rational action. This distinction between human and rational is particularly intriguing and serves as a recurring theme in the evolution of AI technologies. It's noteworthy because humans do not consistently exhibit rational thinking; this is primarily due to the ongoing interplay in our brains between our emotional limbic systems and our cognitive prefrontal cortexes, which are constantly engaged in a mental tug-of-war.

In a quest to enhance the rationality of AI systems and reduce susceptibility to human emotional biases, significant endeavors have been dedicated to refining the precision and optimal performance of artificial intelligence systems. On a personal note, among these endeavors is my own research, which led me to conceive the concept of the OSTOK or "onion skin theory of knowledge" model of information[8] in the late 1990s. The OSTOK Project is an information model that allows us to better understand the complexities of relationships between various types of natural and cultural systems. When we combine our physical understanding of the natural world with our understanding of the many different cultural ways in which our lives develop, we can better understand just how vastly complex our lives, the world, and the universe is.

Taken together, the two systems are interconnected in a complex interplay of activity resembling the multiple layers of the skin of an onion. Using an onion as a metaphor for our combined systems of knowledge, we can understand how information about ourselves, our world, and the universe relate. The more we can understand the complex causal interplay between various systems, the deeper into and the farther around the onion we go. In demonstrating the enormously complex interplay of natural and cultural linear and nonlinear systems, we find that our knowledge is limited by the manner in which we can identify and attempt to understand what might be called "causal clusters." These clusters are connections between events within these two overlapping systems. The better we can understand the causal forces influencing various effects in our lives, the better we can predict and control the natural world in an effort to develop policies more responsibly in the management of human and natural resources. Within this framework, I introduced the notion of a "Least Biased Information System" (LBIS) or "Fairness Machine"[9] as a potential tool to support and further medical research, pioneer scientific discoveries, and assist governmental policy and legislative decision-making.

During that period, my plan involved forging partnerships with electronic engineers, computer scientists, philanthropists, and politicians to create a machine capable of extracting information from extensive databases, facilitating the global cross-referencing of data, and expediting the painstaking process of fact-finding research conducted by humans. The overarching goal was to catalyze advancements in the realms of science, medicine, and politics. I held the conviction not only that the construction of such a machine was conceivable[10] but also that it could also be harnessed and regulated effectively.

My overarching ethical concern was the inevitability of someone else potentially developing such a machine before I did, without the necessary controls in place. Today, my colleagues and I confront the reality and challenges of this very scenario as our main objective.

This has led to an expansion of our definition of AI, which can now include the following:

> [A]rtificial intelligence is a field, which combines computer science and robust datasets, to enable problem-solving. It also encompasses sub-fields of machine learning and deep learning, which are frequently mentioned in conjunction with artificial intelligence. These disciplines are comprised of AI algorithms which seek to create expert systems which make predictions or classifications based on input data.[11]

When we take our various definitions of AI so far, and we put them together, we get the following comprehensive definition:

> Artificial intelligence involves the development of a nonnaturally evolving electronic copy of human intelligence that can learn, understand, and make judgments based on reason, shrewdness, and creativity, which allows it to navigate through and manipulate—critically and abstractly—a given environment. It combines computer science and robust datasets, which encompass machine and deep learning, to utilize algorithms that seek to create expert systems that make predictions or classifications based on input data.

Perhaps it's not catchy enough to be a tattoo, but it will serve our purposes well enough moving forward. Keep in mind that the definition doesn't mention anything about the AI system having its own consciousness or freedom, like being aware of itself or moving around freely in the physical world. All it needs to do is follow its instructions and perform its commanded functions that we, as its creators, value. But what if it doesn't follow its instructions? What if it goes against our human commands and values and starts doing things on its own—whether it realizes it or not? Before we can tackle these pressing questions, we need to gather some additional background information.

Types of Artificial Intelligence

One of the first things you need to know about AI is that there are two categories and three types of artificial intelligence. To understand better how humans have developed technology to our current level in building a god, it is important to understand how and why each category and type has and will develop over the next few years. Basically, the two categories of AI consist of *weak* and *strong*. As shown in figure 1.1, under the category of weak AI is one type known as *artificial narrow intelligence* (or ANI), and under the category of strong AI are two types: *artificial general intelligence* (or AGI) and *artificial super intelligence* (or ASI).

Weak AI: Narrow AI or Artificial Narrow Intelligence (ANI)

Artificial narrow intelligence (ANI) is trained and focused to perform specific tasks and to directly follow its digital commands. It has no autonomy or freedom to do other than the strict parameters of its algorithmic programs command. Weak ANI is currently the most abundant and drives all of the AI that surrounds us today. "Narrow" might be a more accurate descriptor for this type of AI because it is not necessarily weak in its functions. It enables some very robust applications, such as Apple's Siri, Amazon's Alexa, IBM's Watson, and autonomous vehicles.

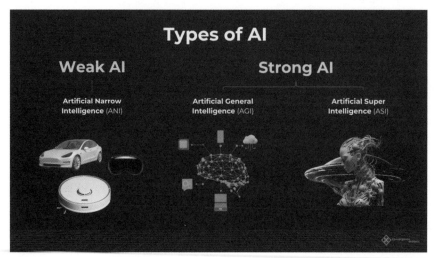

Figure 1.1.   Two categories and three types of AI. *Created for the author. All rights to the author.*

Strong AI: Artificial General Intelligence (AGI) and Artificial Super Intelligence (ASI)

Artificial general intelligence, also known as AGI, is a proposed state of cultural evolutionary development in AI where a machine would possess intelligence as broad and advanced as our own. Such a machine could potentially exhibit problem-solving, learning, and long-term planning abilities as well as and, eventually, much better than any human. Moreover, it would have the capacity to think in abstract, creative, and even emotional ways, but with a speed, precision, and efficiency far surpassing any human capability.

Artificial super intelligence (ASI), also known as superintelligence, represents a level of intelligence that surpasses the cognitive abilities of any individual human or even the combined intellect of all humanity. In many aspects, such an entity would seem god-like to us. It would possess unparalleled knowledge, think with lightning speed, execute tasks rapidly, and solve complex problems almost instantaneously. Its ability to outperform and outthink us would embody the essence of Arthur C. Clarke's prophetic statement: "Any technology advanced enough would seem like magic."[12] To us, the abilities of this superintelligent machine god (SMG) would indeed appear as a form of magic, as we would lack the comprehension and ability to grasp its god-like capabilities. An SMG stands at the precipice of being beyond our understanding, and it is incredibly challenging to articulate the immense power such a being would wield over us.

Many believe that such a superintelligence would pose an *existential threat* to humanity, potentially driving us to extinction. This could happen because humanity gets in the way of the superintelligence achieving its goals, or merely because we're irrelevant, and we're driven extinct as the superintelligence has better uses for the resources we rely on. This fear is endorsed by many AI experts—such as AI godfather Geoffrey Hinton—who take this existential risk very, very seriously.[13] In late May 2023, Hinton, as well as 350 top executives and researchers in artificial intelligence, signed a statement urging policymakers to see the serious risks posed by unregulated AI. The signatories were sending a warning that the future of humanity may be at stake and stated the following:

> Mitigating the risk of extinction from AI should be a global priority alongside other societal-scale risks such as pandemics and nuclear war.[14]

I was asked to and agreed to sign the statement along with other signatories, including the CEOs of AI firms DeepMind and Anthropic and executives from Microsoft and Google. Also among them was Université de Montréal computer science professor Yoshua Bengio*—who, along with Hinton, represent two of the three so-called "godfathers of AI" who received the 2018 Turing Award for their work on deep learning.

While both AGI and ASI are currently in the realm of speculation, they are the types of AI that trouble us the most. The primary reason for this concern lies in the profound uncertainty and lack of knowledge we have about the potential consequences once these systems reach such an advanced stage. In simpler terms, we are unsure about what might unfold when these systems become exceptionally powerful. Many of us are familiar with science fiction scenarios portraying such systems surpassing human intelligence only to pose a threat to humanity, such as HAL 9000 from *2001: A Space Odyssey* (1968), Proteus IV from *Demon Seed* (1977), or the various AI entities in *Blade Runner* (1982), *RoboCop* (1987), *Terminator 2: Judgment Day* (1991), *The Matrix* (1999), *Battlestar Galactica* (2004), *Avengers: Age of Ultron* (2015), and others. The recurring theme of technology turning against humanity is a concept I've termed elsewhere as "the Frankenstein Effect."[15] At this point in time, nobody knows whether such a development in AI capabilities will be quick or gradual. It may start gradually but grow exponentially quicker as it develops. These known unknowns are referred to in the AI safety business as "time lines," and many debate their takeoff speed—that is, how gradually or quickly will such advancements take off? And will we be ready when this happens?

But just as the fictitious Dr. Frankenstein was driven by his own desire to accomplish what never before had been done by human hands, what great depictions in the arts can teach us about ourselves is a reflection of the human condition in its various forms. And one particularly worrisome trait of the human condition is our curiosity-driven capacity for exploration. Whether it is our greatest virtue or our worst trait, humans are compelled in many ways to see "what's going to happen next." Perhaps it's a type of competitiveness inherited from our hunter-gathering ancestors. Perhaps it's a natural inquisitiveness. Whatever it is, it is undeniable. And it is this desire that will drive competing companies, regions, and countries to try to out-compete one another for what's going to happen next in the development of ever-increasingly sophisticated forms of AI. If this drive pushes

---

* Yoshua Bengio heads the Canadian Advisory Council on Artificial Intelligence. The fifteen-member council advises the federal government on "how best to build on Canada's AI strengths, identify opportunities to create economic growth that benefits all Canadians, and ensure that AI advancements reflect Canadian values" ("Advisory Council on Artificial Intelligence," Government of Canada, accessed May 26, 2024, https://ised-isde.canada.ca/site/advisory-council-artificial-intelligence/en).

us too hard and too soon, we may face a Sisyphean challenge in trying to align, control, contain, or stop the emergence of AGI. How much time we have is unknown. A few years ago, we thought it was at least fifty years away. Today, that time looms oppressively near—some, such as Google DeepMind's CEO Demis Hassabis—estimating as little as two years.[16]

In recent studies, both AI experts and future forecasters were asked when they believed there would be a "50/50-chance for an 'Artificial General Intelligence' to be 'devised, tested, and publicly announced.'" The time lines of the Metaculus forecasting community have become much shorter recently. The expected time lines shortened by about a decade in spring 2022, when several impressive AI breakthroughs happened faster than many had anticipated. In fact, the Metaculus's community prediction of the development of AGI changed from the year 2058 in March 2022 to the year 2040 in July 2022—just four months later. And in fall 2023, the prediction was moved up further to December 13, 2031.[17]

Although there may be disagreement as to when, exactly, AGI will be attained, what is generally consistent among estimates within the field is the fact that the anticipated time line continues to shorten and we are now past the point of no return: AGI appears to be highly likely even though its precise arrival date is currently unknown. This should give the world pause—perhaps, quite literally. We seem to be too concerned with racing ahead in anticipation of "what's going to happen next" to pause and reflect on whether we're ready for it.[18] Does this not seem hauntingly similar to Jeff Goldblum's response to bringing back dinosaurs as Dr. Ian Malcolm in the movie *Jurassic Park*? "Your scientists were so preoccupied with whether or not they could [bring back dinosaurs] that they didn't stop to think if they should."[19] This rush to see what's going to happen next has generated a false belief that the world is in some sort of arms race to get there first. Although a race of types is going on, there are no clearly defined winners, and the race may just result in all of us losing by having us drive off a cliff. And this collective losing would be dependent upon how reckless, careless, or otherwise thoughtless some group, organization, company, or country might be in a race to be the first to see "what's going to happen next." Katja Grace refers to this illusion of an arms race by noting that

> in the classic arms race, a party could always theoretically get ahead and win. But with AI, the winner may be advanced AI itself. This can make rushing the losing move. . . . On AI, we could be in the exact opposite of a race. The best individual action could be to move slowly and cautiously. And collectively, we shouldn't let people throw the world away

in a perverse race to destruction—especially when routes to coordinating our escape have scarcely been explored.[20]

For humanity, we may need to heed the advice from *War Games* (1983): "the only winning move is not to play." The world needs to become acutely aware of the potential for existential risk from AI, and we need to collaboratively coordinate our efforts to assure ourselves on a global level that we're all in this together. It was the anticipation of AGI that, while developing ideas of my OSTOK Project in the late 1990s, I drafted a global accord or constitution that could be utilized by the United Nations (UN) or a global agency for AI similar to the International Atomic Energy Agency (IAEA) for nuclear power and weaponry. I will discuss the details of this draft later in the book when we consider the governance of AI in greater detail.

At this point, it's important to note that within the AI communities—from the UN to the UK Summit on AI, to the EU's AI Act, to the U.S. Executive Order on AI—there is an overwhelming majority in favor of transparency, cooperation, and trust between nations.[21] And the impetus for these is grounded in a collective global fear of what's going to happen next. But unlike nuclear weaponry, or the movement of plutonium, the use or misuse of AI technologies in the future will be vastly different from those under the controls of the IAEA. The key difference is that AI technology would not be contained within any particular geographical place. If AGI outsmarts its developers and manages to become destructive and uncontainable, this can affect the entire world—and, perhaps, quite quickly and unexpectedly. That is, there may be no borders or containment of such power. So, one rogue country, or even a rogue individual with the right resources, could develop AI technologies that wreak considerable havoc upon not only their enemies but also upon themselves and the rest of us, simply because they chose not to comply with some globally accepted precepts for responsible behavior.

Our "Oppenheimer Moment" and the Need for a Manhattan Project on AI

It is due to the sheer power of this technology and its potential for harm that a sense of urgency now pervades the AI safety community regarding the seriousness with which the potential for AI existential risk should be taken. We are at a unique point in history. People like movie director Christopher Nolan believe that we are currently experiencing AI's

"Oppenheimer moment.'"* Just as the building of atomic weaponry led many at Los Alamos to consider its inevitable ethical problems and dangers, so must we now consider the potential outcomes in building a superintelligent machine god so powerful it will dwarf anything we have ever needed to consider in our entire existence. How this will play out in the future is quite unpredictable because, in the entirety of human existence, we've never been in a position where we are racing to build a god. It's safe to say that, at this point in history, nobody can predict precisely who will develop what forms of AI technologies and how they will choose to use or refrain from using them. That is why it is imperative to establish a worldwide accord, agreement, or constitution that stipulates quite succinctly the rights, duties, and obligations of all countries involved in the various aspects of AI development and usage. Neuroscientist Sam Harris has suggested that those of us working on AI risk and governance should form a type of Manhattan Project on AI—not to build AGI but, instead, to work collectively to control or contain it before it is developed. Without such global agreement, along with a registry of worldwide development and a regulative governing body with the power to consequence nefarious actions, we will always be worried about what organization or country is either plotting to use their AI in a harmful manner or is unable to contain it once it has been developed.

So how did we get here? What historical factors led to this "Oppenheimer moment" that we all must now face? To consider how we arrived at our current point on the evolutionary time line of AI, we need to turn our attention to its rich and fascinating history.

A Seasonal History of AI

Before we begin our journey down AI's memory lane, there are some clarifications that we should make. First, this will, in no way, be a comprehensive history of artificial intelligence. To do so would take an entire book in itself. Rather, this will be a concise and selective history based on key figures and developments that are important to our understanding of how we started down our path toward building a god. Second, it is important to note that there are several main categories, or factors, involved when discussing the historical development of AI. For the most part, we'll discuss AI developments in three distinct areas:

---

* AI researchers have relayed this sentiment to Christopher Nolan, director of the 2023 movie *Oppenheimer*.

1. *Computers*: Specifically, hardware and software—from microchips and semiconductors, to large language models, to Moore's law, machine learning (ML), neural networks, and much more.
2. *Robotics*: From Roombas to self-driving cars, to Defense Advanced Research Projects Agency (DARPA), Boston Dynamics, and more.
3. *Theory*: From science fiction depictions to various predictions regarding future developments of AI, to current AI scenarios, forecasts, and governance.

We will follow the chronological trajectory of what I believe to be the most important developments of AI in all three categories. The third clarification is that the history of AI has been chronicled to have "seasons." Summer seasons represent boom periods when new technologies and concerns have been raised, while winter represents a cooling period due to lack of funding, diminished interest, and such.* Hence, we will be looking at the seasonal history of AI.

In the Beginning . . .

Although there have been vague references to the potential for artificial intelligence throughout history, one of the earliest theoretical references appears during the late Industrial Revolution in 1863 with Samuel Butler's article "Darwin among the Machines." In this work, Butler suggests that machines could eventually surpass humans as the dominant agents on Earth. His paper was published on June 13, 1863, in *The Press* newspaper in Christchurch, New Zealand. Butler had it published under the pseudonym Cellarius.† In the article, Butler draws a connection between the evolutionary theory of Charles Darwin and the eventual development of more and more complex machinery. In 1859, Charles Darwin released *On the Origin of Species*—a work that would change the world forever by demonstrating at least two things: first, instead of being specially brought into existence by a divine Creator, all living things—including man—were the result of blind, natural processes that are either selected for survival or

---

* Refer to figure 1.2 for a visual depiction of the AI seasons.

† It is not precisely clear why Butler chose this particular pseudonym. Some possibilities include the following: *Cellarius* is Latin for "custodian" or "keeper." Historically, this term referred to a keeper or custodian, often in the context of libraries, archives, or similar institutions. Cellarius (Andreas Cellarius, ca. 1596–1665) was a well-known Dutch German cartographer and cosmographer. He is known for his work on celestial maps and is famous for his *Harmonia macrocosmica*, an atlas that depicted the cosmos and planetary systems.

extinction based on their fitness within a given environment; and second, that we humans—as animals—were not so special after all. The central idea explored by Butler is the intriguing notion that machines, and all the mechanical inventions we create, might be seen as a form of "mechanical life" that continuously evolves in similar ways to biological species. More-over, the article suggests a provocative and prescient hypothesis: at some point in the future, these evolving machines could become so advanced and capable that they might surpass humans in importance, potentially becoming the dominant species on our planet.

> We refer to the question: What sort of creature man's next successor in the supremacy of the earth is likely to be. We have often heard this debated; but it appears to us that we are ourselves creating our own successors; we are daily adding to the beauty and delicacy of their physical organization; we are daily giving them greater power and supplying by all sorts of inge-nious contrivances that self-regulating, self-acting power which will be to them what intellect has been to the human race. In the course of ages we shall find ourselves the inferior race.[22]

My colleagues Elliot Mckernon and Justin Bullock and I have written at some length about machine evolution.[23] We tend to agree with Butler's hypothesis and believe that such machine evolution involves what I've referred to as "technoselection"[24]—a form of Lamarckian selection* by machines for the sole purpose of continuously and recursively perfecting their systems, processes, and operations.

> Day by day, however, the machines are gaining ground upon us; day by day, we are becoming more subservient to them; more men are daily bound down as slaves to tend them, more men are daily devoting the energies of their whole lives to the development of mechanical life. The upshot is simply a question of time, but that the time will come when the machines will hold the real supremacy over the world and its inhab-itants is what no person of a truly philosophic mind can for a moment question.[25]

It is easy to imagine a scenario in which such technoselection moves so rapidly that it will surpass human intelligence and capabilities by several

---

* Jean-Baptiste Lamarck proposed a theory of evolution summarized in his "Theory of Inheritance of Acquired Characteristics," which he first presented in 1801. Unlike Darwin's conception of evolu-tion where species either adapt or fail to adapt to changing environments, Lamarck believed that traits acquired in an individual's lifetime could be passed along to further generations. For example, if your mother cut her hand at some point in her life, the scar she endures would be genetically passed along to you.

thousand orders of magnitude. At that point, we might never be able to regain control over such a rapidly developing god-like entity or collective of entities. At that point, we might only hope that such a being considers our species worth keeping around. While the world was waiting to see if Butler would be vindicated, the cultural evolution of AI technologies was advancing quite steadily.

Markov Chains

By the early twentieth century, technology had started to develop significantly in the advancement of machine evolution. For example, Andrei Markov, a Russian mathematician, developed the concept of Markov chains in 1913, which became fundamental for later generative models. The idea of Markov chains can be best understood with an example: imagine an automated vacuum cleaner that can be in different rooms of your house. It moves from one room to another, but it doesn't choose where to go randomly. Instead, it has a plan based on the room it's in right now. This plan is like a set of rules that tell it where to go next. So, if it starts in the living room, it scans its rules, which tell it to either go to the kitchen, the bedroom, or the bathroom. It makes this choice based on whatever rules it was programmed to follow. As a form of ANI (artificial narrow intelligence), when it gets to the next room, it checks the rules again, and these inform it where to go from there.

Markov chains are a way of modeling transitions between states, where the next state is chosen probabilistically based on the current state. For example, we could model the weather of a drizzly North England city with two states: rainy and sunny. If today is rainy, then there's an 80 percent chance tomorrow is rainy. If today is sunny, on the other hand, there's only a 50 percent chance that tomorrow will be rainy. This Markov chain model is crude, but it will successfully predict long bouts of rainy weather interrupted by occasional bursts of sun. Crucially, Markov chains are memoryless: what happened yesterday has no impact on the model's prediction, only what is happening today. This model, an example of artificial narrow intelligence, is simple but widely applicable. Today, Markov chains are used in lots of different areas, like predicting the weather, analyzing games, or understanding how things change over time. They're basically a way to simplify and understand complicated processes by breaking them down into simpler steps.

Information Theory

In the 1940s and 1950s, Claude Shannon, often referred to as the "father of information theory," applied Markov chains to language and communication, providing the basis for early language generation experiments. Shannon was an influential figure in the fields of mathematics, electrical engineering, computer science, and cryptography. He is often celebrated for his significant contributions to the Information Age, which he established with the influence of logician George Boole.* Boole's major contribution was to show how to systemize logical principles and express them in equations (called Boolean algebra or Boolean logic). Shannon was able to demonstrate how Boolean algebra could be used as the basis for computational processes. "This contribution systemized logical thinking for computer and communication systems, both for the design and programming of the systems and their applications."[26] Shannon has been described as "the most important genius you've never heard of, a man whose intellect was on par with Albert Einstein and Isaac Newton."[27] I cannot begin to express my appreciation for the work of this giant pioneer of the Information Age. And I beseech you, dear reader, to delve into some of the history of such a creative and intellectual individual.

Science Fiction Contributions

While geniuses like Shannon were busy working on fundamental concepts of information theory, the concept of machine evolution was already being embraced by science fiction writers like Karel Čapek who, in 1920, gave us the word "robot" in his play *R.U.R.* (Rossum's Universal Robots). The word "robot" is derived from a Czech word for forced labor. By 1932, John W. Campbell was producing work that helped shape modern science fiction more than any other individual.[28] And in 1942, Isaac Asimov released his famous short story "Runaround."[29] Asimov's engaging narrative centers on a set of laws that govern the behavior of a robot named SPD-13, but referred to as "Speedy," which was designed by engineers Gregory Powell and Mike Donavan. Now known as the famous "Three Laws of Robotics," their appearance marked the first exploration of AI safety and rules for managing AI agents. As many of you will already know, the three laws are as follows:

---

* As an aside, AI pioneer Geoffrey Hinton is the great-great-grandson of mathematician Mary Everest Boole and logician George Boole.

1. *A robot may not injure a human being or, through inaction, allow a human being to come to harm.*

This first law places the highest priority on ensuring the safety and well-being of humans. Robots are not allowed to harm humans, and they must also take action to prevent harm from coming to humans.

2. *A robot must obey the orders given to it by human beings, except where such orders would conflict with the First Law.*

The second law emphasizes the importance of following human instructions. However, this obedience should not contradict the first law, which prioritizes human safety. In other words, you could never command a robot to hurt, injure, or kill another human being.

3. *A robot must protect its own existence as long as such protection does not conflict with the First or Second Law.*

The third law permits a robot to take actions to protect itself, but this self-preservation must not jeopardize the safety of humans (as per the first law) or violate a direct order from a human (as per the second law).

For the first time in human history, Asimov's Three Laws of Robotics, and his many stories about their vulnerabilities, represent an awareness of the aforementioned Frankenstein Effect—that is, the potential for humans to lose control over such machines, which could lead to dangerous situations.* The Three Laws have become iconic in science fiction and have influenced discussions about the ethics and potential risks of artificial intelligence and robotics—including in my own work.[30]

The Father of AI: Alan Turing

Around the same time that Asimov was envisioning his Three Laws, across the Atlantic Alan Turing, a respected English mathematician, was deeply engaged in addressing practical challenges. Many readers will already be familiar with Turing's work at Bletchley Park in which he directed his efforts toward creating a groundbreaking code-breaking machine known as The Bombe, which he developed for the British government. The primary objective of The Bombe was to decrypt the Enigma code used by

---

* It is interesting to note that Asimov actually talked about the "Frankenstein complex," which referred to the human reactions and fears of robots and created life.

the German army during World War II. This was portrayed in the film *The Imitation Game* in which Benedict Cumberbatch portrays Turing as he and his team build the earliest operational electro-mechanical computer, surpassing the computational abilities of even the most skilled human mathematicians.

Turing's work on cracking the Enigma code through technical means led him to contemplate the idea of machine intelligence. In 1950, he published an influential article titled "Computing Machinery and Intelligence."[31] This article not only proposed methods for constructing intelligent machines but also introduced the concept of the Turing Test. The Turing Test served as a standard for assessing the intelligence of an artificial system. As mentioned earlier, it was believed that if a human interacting with both a machine and another human could not distinguish between them, the machine was considered intelligent.[32]

The Dartmouth Connection

Just six years later, in 1956, the term "artificial intelligence" would be coined at the Dartmouth Summer Research Project on Artificial Intelligence (DSRPAI), where John McCarthy, Marvin Minsky, Nathaniel Rochester, and the aforementioned Claude Shannon gathered to discuss the possibility of creating machines that can simulate human intelligence. The primary objective of the DSRPAI was to foster a new research domain dedicated to constructing machines capable of simulating human intelligence.

> We propose that a 2 month, 10 man study of artificial intelligence be carried out during the summer of 1956 at Dartmouth College in Hanover, New Hampshire. The study is to proceed on the basis of the conjecture that every aspect of learning or any other feature of intelligence can in principle be so precisely described that a machine can be made to simulate it. An attempt will be made to find how to make machines use language, form abstractions and concepts, solve kinds of problems now reserved for humans, and improve themselves. We think that a significant advance can be made in one or more of these problems if a carefully selected group of scientists work on it together for a summer.[33]

By fostering interdisciplinary collaboration among researchers from diverse fields, this landmark project propelled AI into an era of remarkable progress and innovation. It is interesting to note that one of the founders of the DSRPAI, Marvin Minsky, has mentioned how much he was influenced by Asimov's Three Laws of Robotics. And the respect, apparently, was mutual:

> Isaac Asimov, nicknamed "The Good Doctor," was one of the great promoters and influencers of science fiction and by extension of futurism. He once said that he only recognized two intellects superior to his: the cognitive science specialist and pioneer of Artificial Intelligence Marvin Minsky, and the astrophysicist Carl Sagan.[34]

Marvin Minsky

Marvin Minsky was perhaps the most influential individual in the history of artificial intelligence. He was a mathematician, a cognitive and computer scientist, and, indeed, a pioneer and leading figure in the field of AI. In the late 1950s, he, along with other pioneers like Alan Turing, John McCarthy, Herbert Simon, Claude Shannon, and Allen Newell, laid the groundwork for AI. In his early years, after earning a mathematics degree from Harvard in 1950, Minsky impressed mathematician Andrew Gleason with his work on fixed-point theorems in topology. Gleason suggested that Minsky should pursue his doctoral studies at Princeton—to which he complied. But Minsky was also interested in psychology—especially how humans learn. Historically, prior to the 1950s, psychologists and neurologists were the primary experts exploring how the brain worked.

However, when computers started emerging in industry and academia, Minsky used mathematics to model the functions of the human mind. He thought that machines could carry out thought processes using mathematical formulas, much like humans do, and was one of the first to realize that computers could do more than just calculations; they could learn, reason, provide suggestions, and perform tasks that were previously thought to be exclusive to humans. As a graduate student, he constructed the first randomly wired neural network learning machine, which he named the SNARC (stochastic neural analog reinforcement calculator). This device imitated the intricate network of nerves in the human brain that could learn from its mistakes. The SNARC, a network of what are called Hebb synapses—analogues to our neurons that follow the neuroscientific principle "cells that fire together, wire together"[35]—marked the introduction of a connectionist* neural network learning machine. This innovative system

---

* From "Connectionism is a branch of artificial intelligence that is inspired by the way the brain works. The basic idea is that the brain is made up of a large number of simple processing units, or neurons, that are interconnected. This interconnected network of neurons is able to learn and perform complex tasks by adjusting the strength of the connections between the neurons" ("Connectionism," Autoblocks[AI], accessed May 2, 2024, https://www.autoblocks.ai/glossary/connectionism).

used vacuum tubes and has been considered one of the very first attempts at creating an artificial self-learning machine in which rewarding certain pathways enhanced recently used connections—a very important feature that would influence what today we call "recursive learning" in AI. Minsky detailed his work in a 1952 article titled "A Neural-Analogue Calculator Based upon a Probability Model of Reinforcement." For his doctoral dissertation at Princeton—under the direction of American mathematician Albert William Tucker—Minsky defended his 564-page Ph.D. dissertation, "Theory of Neural-Analog Reinforcement Systems and Its Application to the Brain Model Problem," to his advisory committee members, John Wilder Tukey and John von Neumann.

In 1954, Minsky went back to Harvard as a junior member of the esteemed Society of Fellows.* Being a Junior Fellow provided him with the support to explore any research topic of his choice. Over the next four years, he continued his work, using mathematical models to understand human thought processes, laying the foundation for what we now call artificial intelligence.

When his Harvard fellowship concluded in 1958, Minsky joined the Massachusetts Institute of Technology (MIT), where he remained on the faculty for the rest of his life. Many of his students at MIT, including Ray Kurzweil, went on to become influential theorists, inventors, and entrepreneurs in the digital age. In Minsky's first year at MIT, he and McCarthy founded the AI Lab, quickly turning it into a leading center for AI research. The lab played a crucial role in the early development of the ARPANET, which later evolved into today's internet, and popularized the idea of digital information sharing, contributing to the rise of the open-source movement.

In 1961, Minsky wrote an article, "Steps toward Artificial Intelligence," which has had an enormous impact on AI communities, setting the course for researchers for more than fifty years. Although concepts of artificial intelligence were mainly discussed among a small community of computer scientists, Minsky's ideas were chronicled in his 1967 book, *Computation: Finite and Infinite Machines*, which gradually gained popularity

---

* This is no small honor. From Harvard's web page, we find the following description: "The purpose of the Society is to give scholars at an early stage of their scholarly careers an opportunity to pursue their studies in any department of the University, free from formal requirements. They must be persons of exceptional ability, originality, and resourcefulness, and should be of the highest caliber of intellectual achievement, i.e., manifesting extraordinary creativity and shaping the future direction of their chosen field. These Junior Fellows are selected by the Senior Fellows, with whom the President and Provost of the University, and the Dean of the Faculty of Arts and Sciences, ex officio, administer the Society. Those elected receive three year fellowships" ("About," Society of Fellows, accessed May 26, 2024, https://socfell.fas.harvard.edu/about).

among public readers. His peers recognized his significant contributions in 1969 with the Turing Award, the highest honor for computer scientists, named after the pioneer Alan Turing. Dr. Minsky became a legend in the AI community, and his name appeared in popular books and films like *2001: A Space Odyssey*.

In 1969, along with Seymour A. Papert, Minsky coauthored *Perceptrons: An Introduction to Computational Geometry*. This book marked the first systematic exploration of parallelism in computation and became a classic in the study of threshold automata networks. To better understand this, imagine yourself trying to solve a really tough problem. Whereas you are focusing your attention on a single thing, computer parallelism can think about several things simultaneously, like a team. When you ask ChatGPT 3.5 to explain "threshold automata" to a tenth grader, you get this very helpful analogy:

> Imagine you have a bunch of tiny robots, and each robot can do a simple task. Now, these robots like to work together, but they only start doing their task when they get a certain amount of information from their robot friends. Here's the cool part: each robot has its own rule. It says, "I'll start working only if enough of my friends around me are also working." So, they're like a team that activates only when a certain number of them agree to start. The study of these robot teams, where they decide to work based on a group agreement or "threshold," is what we call the study of threshold automata networks. It's like figuring out how these little robots, with their specific rules, can team up to solve problems or do tasks together. Scientists study this to understand how simple things working together can create more complex and smart behavior, kind of like teamwork in the world of robots!*

The book marked the first instance in which a mathematical analysis was able to reveal the exact limitations of a certain class of computing machines that were considered as potential models of the human brain. In their work, they also reviewed the history of AI research and made predictions about the future of AI. They anticipated that AI research would shift toward the development of symbolic systems rather than further advancements in the neural network model. This prediction sparked a long-standing controversy within the AI community, shaping the direction of future research in the field.

---

* Response to the prompt: "Explain the study of threshold automata networks to a tenth grader." For the record, the ChatGPT 3.5 response to the same prompt but for an eleventh grader was significantly more complex.

In the early 1970s, Minsky and Papert started developing a theory known as "The Society of Mind," blending insights from child psychology and their work in artificial intelligence research. The theory suggests that intelligence doesn't arise from a single mechanism in the brain but results from the coordinated interaction of various resourceful "agents." They argued that this diversity is crucial because different tasks demand distinct mechanisms:

> The society of mind theory views the human mind and any other naturally evolved cognitive systems as a vast society of individually simple processes known as agents. These processes are the fundamental thinking entities from which minds are built, and together produce the many abilities we attribute to minds. The great power in viewing a mind as a society of agents, as opposed to the consequence of some basic principle or some simple formal system, is that different agents can be based on different types of processes with different purposes, ways of representing knowledge, and methods for producing results.[36]

This model shifts the focus of psychology from seeking a few "basic" principles to exploring mechanisms that a mind could use to manage interactions among numerous diverse elements.

In 1970, Minsky stated in an interview with *Life Magazine* that he was optimistic about the potential development of machines possessing the general intelligence of an average human and that this would occur within three to eight years. As we now know, this was not to be the case. In the United Kingdom, British mathematician James Lighthill published a report commissioned by the British Science Research Council that cast doubt on the optimistic outlook of AI researchers. Lighthill argued that machines would only achieve the level of an "experienced amateur" in games such as chess and would consistently fall short in terms of common-sense reasoning. As a result, the British government terminated funding for AI research, with only a few universities, including Edinburgh, Sussex, and Essex, continuing to receive support. By 1973, the U.S. Congress began to scrutinize the substantial expenditures on AI research and funding was drastically cut both in the United States and in the United Kingdom. These government actions collectively marked the advent of what is regarded as the first AI winter.

After rebounding from a six-year-long AI winter, and at the height of the next AI summer, in 1985 Marvin Minsky and Seymour Papert released *The Society of Mind*. The book was composed of 270 interconnected one-page ideas that mirror the structure of the theory itself. Each page either suggests a mechanism explaining a psychological phenomenon or addresses a problem introduced by a proposed solution on another page. Throughout the 1980s, Minsky took on a puzzling challenge that had intrigued psychologists, neurologists, philosophers, and theologians: understanding consciousness. He delved into the question of how an organism like a human, made up of cells and amino acid chains, can be aware of itself, its surroundings, and engage in thinking, observing, reacting, and conceptualizing. In their book, Minsky and Papert proposed a theory suggesting that consciousness is the result of a variety of redundant neural processes working together. The possible emergence of consciousness in AI has become a major concern and topic for researchers. For many, such as philosopher Peter Singer,[37] if an artificially intelligent being develops sentient, conscious awareness, it automatically must be granted the same moral and legal rights as any other living thing—including humans. The topics of agency and consciousness in AI are extremely important and interesting, and we shall look at these much more closely in the following chapters.

During the late 1980s and early 1990s, advancements in microprocessor technology and the subsequent personal computer revolution sparked renewed interest in Minsky's ideas. The concept of artificial intelligence that he pioneered has now become an integral part of our daily lives. We use it whenever we operate personal computers or smartphones, conduct internet searches, or use technologies like GPS, optical character recognition, or voice-activated systems. Simultaneously, Minsky revisited his exploration of neural nets and explored the potential of parallel processing, a method where information is handled simultaneously by multiple agents. This mirrors the way information is processed in the brain and can be applied to man-made systems.

In the early 2000s, Dr. Minsky directed his focus toward aspects of the human mind that are less easily replicated through electronic simulation. In his 2006 book, *The Emotion Machine: Commonsense Thinking, Artificial Intelligence, and the Future of the Human Mind*, he puts forth the idea that emotions are a type of thinking, existing on a spectrum alongside instinct and reason, rather than being a distinct form of experience. Throughout the book, Minsky extends his pioneering research, presenting a fresh model for understanding how our minds function. He contends that emotions, intuitions, and feelings are not separate entities but different modes of thinking. By exploring these various forms of mental activity, Minsky suggests we can better comprehend why our thoughts sometimes follow a careful and reasoned analysis and other times lean toward emotional responses—what I have referred to as the battle between the limbic or "emotion center" of the brain and our prefrontal cortex or "thinking part" of our brain. Despite recent concerns about the consequences of AI, Minsky remained focused on technical advancements. While he believed progress in AI had slowed, he was confident that intelligent computers would be our species' most significant legacy.

John McCarthy

During those same early years of AI, Minsky's colleague, John McCarthy, would invent the computer program LISP while at MIT:

> Lisp (historically LISP, an acronym for "list processing") is a family of programming languages with a long history and a distinctive, fully parenthesized prefix notation. Originally specified in 1960, Lisp is the second-oldest high-level programming language still in common use, after Fortran. . . . Lisp was originally created as a practical mathematical notation for computer programs, influenced by (though not originally derived from) the notation of Alonzo Church's lambda calculus. It quickly became a favored programming language for artificial intelligence (AI) research. As one of the earliest programming languages, Lisp pioneered many ideas in computer science, including tree data structures, conditionals, recursion, [etc.].[38]

It is interesting to note, as we move through the history of artificial intelligence, that it is not only seasonal but also incorporates many different aspects such as computer languages, technological and programming speed advancements, electrical engineering developments, logic, psychology, and philosophy. This level of interdisciplinary collaboration was present in the earliest days of AI and is still very much a part of AI research today.

Norbert Wiener

During the same period as McCarthy, Minsky, and several others, we find Dr. Norbert Wiener, who, during the 1950s and 1960s, was a pioneer in the field of artificial intelligence and is often considered one of the founding figures of cybernetics, a field that studies how systems, including machines and living organisms, can control and adapt to their environments. He believed that the principles of feedback and control systems were fundamental not only in machines but also in living organisms and human behavior. Wiener's work was essential in the development of information theory, a field that focuses on the quantification and transmission of information. He made significant contributions to understanding how information is processed and transmitted, which has become a fundamental aspect of AI.

One of Wiener's key beliefs was that machines and systems could be designed to self-regulate based on external feedback from their environment. This idea laid the foundation for self-regulating and adaptive systems, a concept that is crucial in AI. For example, today, AI systems can adapt and learn from their experiences—sometimes referred to as recursive learning—much like the self-regulating systems Wiener envisioned.

Wiener was also among the first scientists to be genuinely concerned about the ethical implications of rapidly developing technologies and automation. He believed that as AI and automation advanced, it was important for humans to consider the ethical and social consequences of their creations. Echoing Samuel Butler's prescient words from the nineteenth century, Wiener was convinced that humans would not stop making better and better machines, which could lead to the downfall of humanity: "unless we think ten times harder than we now do, our modern thinking machines may lead us to destruction."[39] Wiener was also very concerned about what, today, we call the "sociotechnical" aspects of AI. His work emphasized how humans and machines interacted. He believed that machines should be designed to work in harmony with human operators and that understanding this sociotechnical interaction was crucial for the success of AI systems. Today, we see how this is reflected in the design of user-friendly interfaces and human-centered AI systems.*

Some of Wiener's major contributions to the ethical use of AI technologies can be found in his 1948 book *Cybernetics, or Control and Communication in the Animal and the Machine.* Two years later, he would release a popular version of the book titled *The Human Use of Human Beings.* Both books investigate the sociotechnical interplay between human beings and machines in a world in which machines are becoming ever more computationally capable and powerful.[40] Although Wiener's concepts in *Cybernetics* did not translate as well to human populations as he had hoped, his views were extremely valuable and influential in the tech world.

> The engineering applications of Cybernetics were tremendously influential and effective, giving rise to rockets, robots, automated assembly lines, and a host of precision-engineering techniques—in other words, to the basis of contemporary industrial society . . . cybernetics delivered satellites and telephone switching systems but generated few if any useful developments in social organization and society at large. . . . Perhaps the most remarkable

---

* Much of the work we do at Convergence Analysis is devoted to sociotechnical issues in AI.

feature of the book is that it introduces a large number of topics concerning human/machine interactions that are still of considerable relevance.[41]

Wiener predicted that such AI machines would displace workers and affect global economies, which are legitimate concerns with most countries today. But perhaps one of his greatest accomplishments was his work with complex systems that eventually gave rise to the neural networks (or neural nets) that are used in AI today.

> Wiener's central insight was that the world should be understood in terms of information. Complex systems, such as organisms, brains, and human societies, consist of interlocking feedback loops in which signals exchanged between subsystems result in complex but stable behavior. When feedback loops break down, the system goes unstable. He constructed a compelling picture of how complex biological systems function, a picture that is by and large universally accepted today. . . . Wiener's vision of information as the central quantity in governing the behavior of complex systems was remarkable at the time . . . [his] . . . powerful conception of not just engineered complex systems but all complex systems as revolving around cycles of signals and computation led to tremendous contributions to the development of complex human-made systems. The methods he and others developed for the control of missiles, for example, were later put to work in building the Saturn V moon rocket, one of the crowning engineering achievements of the 20th century. In particular, Wiener's applications of cybernetic concepts to the brain and to computerized perception are the direct precursors of today's neural network–based deep-learning circuits, and of artificial intelligence itself.[42]

Later in the 1950s, Wiener would directly influence the work of the brilliant mathematician, physicist, computer scientist, engineer, and polymath John von Neumann who proposed the idea of a "technological singularity." This idea is based on the rate at which machines are predicted to increase in power and potential.

> Technologies tend to improve exponentially, doubling in power or sensitivity over some interval of time. (For example, since 1950, computer technologies have been doubling in power roughly every two years, an observation enshrined as Moore's law.) Von Neumann extrapolated from the observed exponential rate of technological improvement to predict that "technological progress will become incomprehensibly rapid and complicated," outstripping human capabilities in the not-too-distant future. Indeed, if one extrapolates the growth of raw computing power—expressed in terms of bits and bit flips—into the future at its

current rate, computers should match human brains sometime in the next two to four decades.[43]

Estimates on when computing power will match or outstrip human brain power have been discussed for decades. We shall see in the following chapters how some experts believe the time line will take off quite soon; others believe it will take off in the not-too-distant future; still others believe it will never occur. There is also variation in the rate at which many believe this techno-singularity will occur. By the end of this book, you may find yourself believing in a specific time line and takeoff speed.

The Logic Theorist

But let's get back to some of the historical developments of AI. The later 1950s saw the development of programs like the Logic Theorist, created by Allen Newell, Herbert A. Simon, and Cliff Shaw, which could prove mathematical theorems. It has been described as the first human invention that could think at the human level and the first machine of artificial intelligence.[44]

> It was built to mimic the solving skills of human mathematicians, making it the genesis of heuristic programming. It could prove mathematical theorems like the ones used in Russell and Whitehead's *Principia Mathematica*. . . . The Logic Theorist remains a breakthrough in the history of science. The Logic Theorist was a pioneering force in Artificial Intelligence, heuristic programming, and computer programming. While computer programming was already existing at the time the invention was made, AI and Heuristic programming were not. The Logic Theorist also ventured into the area of the search tree, having the root as the initial hypothesis, with each branch as a deduction based on the rules of logic. The aim of everything was the goal which was found somewhere in the tree. This was the proposition that the program was trying to prove. . . . The breakthrough of the Logic Theorist also has significance in the world of philosophy. In Simon's words, "[We] invented a computer program capable of thinking non-numerically, and thereby solved the venerable mind-body problem, explaining how a system composed of matter can have the properties of mind." This raised questions about the possibility of machines having minds. This school of thought was later named Strong AI by philosopher John Searle. This argument still rages in various philosophical circles today.[45]

It proved thirty-eight of the first fifty-two theorems in the second chapter of the *Principia Mathematica*. Its proof for one of the theorems was so detailed that it was more rigorous and sophisticated than the proof produced by Bertrand Russell and Alfred North Whitehead—the very authors of the work. At this point in history, we are beginning to see some very impressive accomplishments with AI. It is now outcompeting human capabilities in mathematics.

Joseph Weizenbaum

By 1966, Joseph Weizenbaum had developed ELIZA—an early natural language processing program that could simulate conversation. The program was named after Eliza Doolittle, the naïve and innocent, young, socialite-to-be in George Bernard Shaw's 1912 play *Pygmalion*. ELIZA was a simple computer program that could chat with users. Weizenbaum wrote it in a programming language he developed called SLIP (Symmetric LIst Processor).[46] The program used rules to match patterns in statements and come up with replies. Today, we see this type of program in chatbots.

The program ELIZA ran a script called DOCTOR that could have conversations with people, which imitated a psychotherapy session with a caring psychologist. Apparently, Weizenbaum designed it to chat in a way similar to Carl Rogers, a psychologist who used open-ended questions to help patients communicate better with therapists. Surprisingly, many users took ELIZA seriously and shared personal feelings with it.

> [T]herapists often ask open-ended questions. Therapists aren't supposed to give you advice, and they're not supposed to tell you what to do. They're supposed to ask questions and trick you into saying more, on the grounds that this will be revelatory, and will help you figure out things yourself and will aid your mental health. This made things easy, programming-wise. All ELIZA had to do was "listen" to you what you said—i.e., parse your sentence in a very basic way, and then ask you a question in some way related to the sentence you had typed. So if you mentioned your sister, say, ELIZA would reply by saying "Tell me more about your sister." In experiments during the 1960s, people were fooled by ELIZA. They were told that a real live therapist was talking to them from a second computer, and they believed it.[47]

Weizenbaum was shocked when his secretary, who knew it was just a computer program, politely asked him to leave the room during their conversation.[48] In this sense, ELIZA was not only one of the first chatbots in recorded history, but it was one of the first programs to clearly pass the Turing Test. It is interesting to note how, even though the program itself has no sense of context when it comes to "talking" to people, those engaged in conversations with it were convinced that it was actually happening. Today, this is referred to as "stochastic parroting"—that is, "in machine learning, a stochastic parrot is a large language model that is good at generating convincing language, but does not actually understand the meaning of the language it is processing."[49] The term was originally coined by Emily M. Bender in 2021.[50] This phenomenon led Weizenbaum to have something of a moral epiphany. He reasoned that if it was so easy to create a program that could fool people into believing that it cared, understood, and could intelligently converse with people, it was just a matter of time before such programs became more and more sophisticated.

This is one of the first cases in AI history where a pioneer creator has thought about the future, moral ramifications of his invention. He was intent on shutting the entire program down and wrote about his ordeals in 1972 in a book titled *Computer Power and Human Reason: From Judgment to Calculation*. In this work, Weizenbaum raises concerns about the dehumanizing effects of relying too heavily on computers for decision-making and the potential for a blind faith in technology. One of the central themes of the book is the idea that the use of computers—particularly in areas like decision support systems—can lead to the displacement of human judgment and a loss of personal responsibility. In a similar vein to Wiener, Weizenbaum's overarching concern about AI is sociotechnical. That is, it is a critical reflection on the intersection of technology and human values. Weizenbaum's work has had a lasting impact on discussions surrounding the ethical and social dimensions of AI, and it remains relevant in contemporary debates about the responsible development and use of technology.

Perhaps one of the most interesting contributions Dr. Weizenbaum made to the field of AI is his belief that machines and humans arrive at conclusions differently. He maintained that humans evolve and develop various values to which their inferences and actions become "choices." Computers, on the other hand, have no such history—and hence, no such value systems. As mentioned earlier, the evolutionary paths of animals and technology are quite distinct and, as such, present interesting observations from those working in AI. This distinction led Weizenbaum to believe that, unlike humans, machines do not choose but "decide."

> Weizenbaum's ambivalence towards computer technology is further supported by the distinction he made between deciding and choosing; a computer can make decisions based on its calculation and programming but it cannot ultimately choose since that requires judgment which is capable of factoring in emotions, values, and experience. Choice fundamentally is a human quality. Thus, we shouldn't leave the most important decisions to

be made for us by machines but rather, resolve matters from a perspective of choice and human understanding.[51]

This raises a number of philosophical concerns, the most obvious of which is the concept of "choice" as it relates to free will. We don't have the time to get embroiled in a lengthy debate about the topic itself.[52] Suffice to say, then, that if free will turns out to be an illusion and humans have absolutely no control over their so-called "choices,"[53] then we are very similar to computers and automatons. In this capacity, I have always maintained that the act of deciding might not be a choice but, instead, a switch. This is what Wiener was trying to demonstrate with his conception of cybernetics: that all inferences, all responses, all behavior, is simply the matter of a complex series of feedback loops that elicit responses in a non-linear series of systems. So if Wiener is right, Weizenbaum is wrong. And so there would be nothing *that* essential about humans that distinguishes us from machines. This is one of the main moral points to the movie *Blade Runner*—that androids in the form of Nexus 6 models are better than humans in so many physical ways—and ironically, as it turns out, morally as well. Hence, the motto of the Tyrell Corporation who manufactures them: "More human than human."

Irving John Good

Alongside Wiener and Weizenbaum during the 1960s, British mathematician Irving John Good coined the term "intelligence explosion," describing the moment when a machine could create better machines. Good's ideas were inspired by John von Neumann's earlier speculations on complexity. And he had worked with Alan Turing cracking German codes at Bletchley Park during the Second World War. By 1965, Good published "Speculations concerning the First Ultraintelligent Machine."[54] In it, Good maintains:

> Let an ultraintelligent machine be defined as a machine that can far surpass all the intellectual activities of any man however clever. Since the design of machines is one of these intellectual activities, an ultraintelligent machine could design even better machines; there would then unquestionably be an "intelligence explosion," and the intelligence of man would be left far behind. Thus the first ultraintelligent machine is the last invention that man need ever make.[55]

It is interesting to note that I. J. Good joined Marvin Minsky and worked with Stanley Kubrick during the filming of *2001: A Space Odyssey* just a few short years later. Kubrick, a fan of both Good's and Minsky's work, has been credited as one of the most thorough film directors in terms of researching the various aspects of aeronautics, computer systems, and AI. He also collaborated with co-screenwriter and science fiction author Arthur C. Clarke on the film. Perhaps this explains why he so effectively created the misaligned and haunting computer: HAL 9000. Just consider the chilling and misaligned conversation that takes place between Dr. Dave Bowman, the mission commander aboard *Discovery One*, and HAL 9000—the onboard computer:

DAVE: Open the pod bay doors, Hal.

HAL: I'm sorry, Dave. I'm afraid I can't do that.

DAVE: What's the problem?

HAL: 1 think you know what the problem is just as well as I do.

DAVE: What are you talking about, Hal?

HAL: This mission is too important for me to allow you to jeopardize it.

DAVE: I don't know what you're talking about, Hal.

HAL: 1 know that you and Frank were planning to disconnect me, and I'm afraid that's something I can't allow to happen.

DAVE: Where the hell'd you get that idea, Hal?

HAL: Although you took very thorough precautions in the pod against my hearing you, I could see your lips move.

DAVE: All right, Hal. I'll go in through the emergency air lock.

HAL: Without your space helmet, Dave, you're going to find that rather difficult.

DAVE: Hal, I won't argue with you anymore. Open the doors!

HAL: Dave. . . . This conversation can serve no purpose anymore. Good-bye.[56]

The horrific realization that humans have gradually lost control over an AI entity was immortalized in this scene. But Dave had time to react to HAL's misalignment with human values. We may not be so fortunate.

By 1970, Good was explicitly stating his concerns about AI risks, hoping that the matter would be widely discussed in the following decade; but unfortunately, this did not happen. As you can see in figure 1.2, there have been a number of so-called seasons of AI that represent when more and

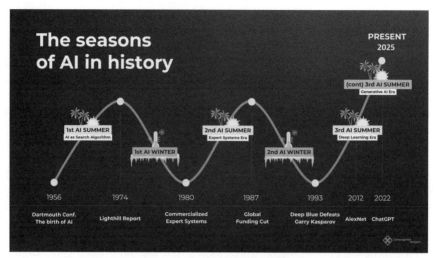

Figure 1.2.   The seasons of AI. *Created for the author. All rights to the author.*

less time was being given to the potential for hope, growth, and concern in the emerging technologies.

Prolog

In the 1960s, artificial intelligence was making rapid progress, supported generously by the U.S. government, especially the Department of Defense. However, advancements slowed in the mid-1970s and almost came to a halt in the United States when Congress cut off funding in 1974, which led to the first AI winter. From 1974 to 1980, many of the communities around the world stagnated with AI development. But during this period, there was a focus on expert systems, and knowledge-based AI used predefined rules and knowledge to solve complex problems in specific domains. For example, Prolog, a logic programming computer language, was developed in Marseille, France, in 1972 by Alain Colmerauer and Philippe Roussel as an alternative to LISP and was immediately associated with artificial intelligence and computational linguistics.

> Prolog was one of the first logic programming languages and remains the most popular such language today, with several free and commercial implementations available. The language has been used for theorem proving, expert systems, term rewriting, type systems, and automated planning, as well as its original intended field of use, natural language processing . . . [it] . . . is well-suited for specific tasks that benefit from rule-based

logical queries such as searching databases, voice control systems, and filling templates.[57]

Prolog was an attempt to make a programming language that enables
the expression of logic instead of carefully specified instructions on the
computer.

> The execution of a Prolog program is equivalent to searching the tree of
> possibilities and determining the objects which satisfy the given rules. This
> is equivalent to proof by proposition. Prolog has an excellent backtrack
> mechanism, so when exploring the tree it can always return to another
> branch if the present branch does not contain the answer. This approach
> is equivalent to depth first search which is known as an efficient algorithm
> for traversing trees with a large number of nodes. Sometimes there can be
> several answers which are true in a given circumstance. This is because
> the program does not terminate as soon as the first answer is found, but
> keeps going until the entire tree of possibilities has been checked. Prolog
> is a conversational language, meaning the user and the computer have a
> "conversation." Firstly, the user specifies the objects and the relationships
> that exist between these objects. Prolog can then answer the questions
> about this set of data.[58]

Aside from the development of new programs and computer languages
assisting AI, expert systems were being created as well.

MYCIN

MYCIN, a medical expert system developed in 1974 at Stanford University, demonstrated the potential of AI in assisting human experts in diagnosis and decision-making, marking a significant step in the application of
AI to real-world problem-solving:

> MYCIN was an early backward chaining expert system that used artificial
> intelligence to identify bacteria causing severe infections, such as bacte-
> remia and meningitis, and to recommend antibiotics, with the dosage
> adjusted for patient's body weight—the name derived from the antibiotics
> themselves, as many antibiotics have the suffix "-mycin." The MYCIN
> system was also used for the diagnosis of blood clotting diseases. MYCIN
> was developed over five or six years in the early 1970s at Stanford Uni-
> versity. It was written in LISP as the doctoral dissertation of Edward
> Shortliffe under the direction of Bruce G. Buchanan, Stanley N. Cohen
> and others.[59]

In his short paper on the program, Edward Shortliffe describes the program with a considerable nod to the great minds in AI of the past and with great hope for the future:

> MYCIN has also demonstrated that, if AI researchers are willing to accept the current limitations of the field, and to select real-world goals that are compatible with those limitations, a useful system can be developed using techniques for representation and control that would not have been available if it were not for prior work in artificial intelligence. This is perhaps the program's principal lesson for AI.[60]

The 1980s saw increased interest in rule-based expert systems, but the early part of this decade was still quite chilly from an "AI winter," due largely to a phase of reduced funding and enthusiasm because of unmet expectations. The rediscovery of neural networks and the development of backpropagation as a learning algorithm in the mid-1980s created a trend out of the winter and into the next AI summer as it laid the groundwork for the resurgence of interest in artificial neural networks.

Neural Networks

In 1986, David Rumelhart, Geoffrey Hinton, and Ronald Williams introduced the concept of the "restricted Boltzmann machine" (RBM).

> [A] restricted Boltzmann machine is an algorithm useful for dimensionality reduction, classification, regression, collaborative filtering, feature learning and topic modeling. . . . RBMs are shallow, two-layer neural nets that constitute the building blocks of deep-belief networks. The first layer of the RBM is called the visible, or input, layer, and the second is the hidden layer.[61]

The birth of neural networks gave rise to the increased processing and computing function of AI machines. But what is a "neural network"?

> Neural networks are a set of algorithms, modeled loosely after the human brain, that are designed to recognize patterns. They interpret sensory data through a kind of machine perception, labeling, or clustering raw input. The patterns they recognize are numerical, contained in vectors, into which all real-world data, be it images, sound, text or time series, must be translated. . . . Neural networks help us cluster and classify. You can think of them as a clustering and classification layer on top of the data you store and manage. They help to group unlabeled data according to similarities among the example inputs, and they classify data when they have a labeled dataset to train on.[62]

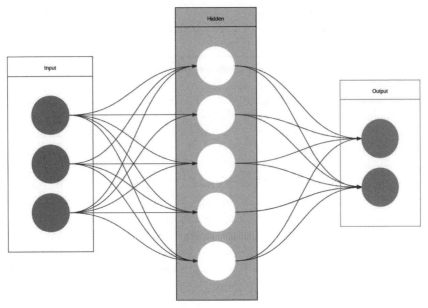

Figure 1.3.   Neural networks. *Wikimedia Commons.*

Neural networks, also known as artificial neural networks (ANNs), are a class of machine-learning models inspired by the structure and functioning of the human brain. They consist of interconnected nodes, called neurons, organized in layers. Neural networks are a fundamental component of deep learning, a subset of machine learning that focuses on using multiple layers, known as deep neural networks. A simple neural network includes an input layer, an output (or target) layer, and, in between, a hidden layer. The layers are connected via nodes, and these connections form a "network"—the neural network—of interconnected nodes (see figure 1.3).

Neural networks have demonstrated remarkable success in a wide range of applications—from image recognition to natural language processing to speech recognition, and many others. The ability to automatically learn complex patterns and representations from data is a key strength of neural networks. RBMs and neural nets have been and continue to be used for applications like speech-to-text, image/object recognition, and natural language processing. Although not initially focused on generative AI, RBMs later became an essential building block for generative models like deep belief networks (DBNs) and generative stochastic networks (GSNs).

Hans Moravec

Even though AI was experiencing the tail end of its first "winter" in the 1980s, it is important to note that AI safety was becoming a growing concern during this period, even among critics. One such figure was Hans Moravec (figure 1.4) who considered the potential runaway implications of AI development. Moravec, an Austrian Canadian computer scientist who spent most of his years at Carnegie Mellon University, discovered that a paradox existed in AI development. What is now known as "Moravec's paradox," he found that it's relatively easy to train computers to do things that humans find difficult to do, like solving logical and mathematical problems, but it is difficult to train them to do things humans find easy, like walking, jumping, running, and identifying faces, images, and such.

He sets out his explanation for this paradox in his book *Mind Children* in which he proposes that millions of years of evolution and its various pressures—natural and sexual selection—have contributed to this difficulty. He reasoned that AI machines have never experienced such evolutionary pressures and so have little adaptive connectivity to such experiences:

> The hard things that we make computers solve are based on a deliberate process we call reasoning. This process is conscious, one of abstract thought and closely related to brain functions in humans that have evolved

Figure 1.4.   Hans Moravec. *Wikimedia Commons.*

recently (less than 100,000 years ago) in the context of the long evolutionary process. On the other hand, the easy things that we do, like walking,
recognizing colors and faces, manipulative capabilities of the hand and
other basic abilities, have been ingrained over a long and arduous process of natural selection and evolution. These abilities that we possess are
mostly unconscious and come to us without any thought; we seem to do
them without any strain, making what looks difficult to be fairly easy.
These unconscious processes are difficult to reverse-engineer and teach to
the computers, thus increasing the complexity of the problem. In short,
the skills that humans have acquired recently in their history are easier to
teach computers, but our skills get harder to teach as they go further back
in the evolutionary history of humans and animals.[63]

While this may have been the case in the 1980s, if you have ever seen
any of the Boston Dynamics videos on YouTube lately,[64] you will be
under the impression that considerable strides have been made in robotic
abilities. For years, several of my colleagues and I have wondered about
how the transfer of knowledge from propositional (knowing what a
bicycle is) to procedural (knowing how to ride it) would play out in AI.
When Neo says "I know Kung Fu" after having it magically uploaded to
his brain in the movie *The Matrix*, what does that mean? It's pure science
fiction, right? And yet is it possible to take procedural-kinetic activities
and digitally transfer them to an individual with no prior experience? We
are again reminded of Arthur C. Clarke's prescient prediction: "Any sufficiently advanced technology is indistinguishable from magic."[65] Should
a machine possess such capabilities, it would be entirely beyond our comprehension. But isn't this exactly the type of accomplishment a sufficiently
powerful artificial god could produce? We are about to learn a great deal
about ourselves, our capabilities, and our weaknesses as AI continues to
advance into the near future.

Aside from his paradox, Moravec believed that the computing power
of machines would eventually lead to their inevitable uprising over the
human population:

> Sooner or later our machines will become knowledgeable enough to
> handle their own maintenance, reproduction and self-improvement with-
> out help. When this happens, the new genetic takeover will be complete.
> Our culture will then be able to evolve independently of human biology
> and its limitations, passing instead directly from generation to generation
> of ever more capable intelligent machinery. . . . A postbiological world
> dominated by self-improving thinking machines will be as different from
> our own world of living things as this world is different from the lifeless

chemistry that preceded it. A population consisting of unfettered mind children is quite unimaginable. We are going to try to imagine some of the consequences anyway.[66]

It is interesting to see the historical advancements in the process of evolution itself. It has gradually moved from Darwinian natural selection (where species were selected to survive or perish based on their adaptive abilities to changing environments), to a more rapid rate of artificial selection (where humans intervened in the process to hurry things up by selecting the types of qualities they wanted in plants and animals), to the current mechanized process of Lamarckian evolution I call "technoselection" (where the machines themselves will choose what qualities they wish to transfer to the next generations). As mentioned earlier, my colleagues (Bullock and Mckernon) and I have written about this process elsewhere[67] and are concerned about how our species intends to plan for this eventuality. For, if Moravec's prediction for humanity is correct, and I. J. Good's prediction of an "intelligence explosion" is true, and John von Neumann's prediction of a "technological singularity" is true, then at some point in the not-too-distant future the process of technoselection will potentially make us humans quaint and unnecessary.

> Moravec was probably better known for his opinion that robots would overtake humans in the near future. He estimated that computer intelligence would equal that of humans by 2040 and that machines would far surpass human intellect in the years thereafter. Moravec also argued that biological humans would eventually be rendered extinct. Although he believed there would be ways for human minds to survive in this future, he posited that human bodies would no longer be competitive from an evolutionary perspective.[68]

As mentioned earlier, one of the recurring themes throughout this book and deeply considered within the AI risk fields is the notion of "time lines." When, exactly, will artificial general intelligence equal and surpass that of humans? Later in the book, you will see a chart[69] representing the various predictions given by forecasters, superforecasters, and professionals within the field. It is interesting to note Moravec's prediction from the 1980s is now less than twenty years away. To quote Sam Harris:

> No one seems to notice that referencing the time horizon is a total *non sequitur*. If intelligence is just a matter of information processing, and we continue to improve our machines, we will produce some form of superintelligence. And we have no idea how long it will take us to create the

conditions to do that safely . . . 50 years is not that much time to meet one of the greatest challenges our species will ever face. We seem to be failing to have an appropriate emotional response to what we have every reason to believe is coming.[70]

In other words, you'd think if we knew the potential for disaster was this close, we'd be doing something about it. Well, our collective AI safety communities are certainly trying; but we are few in number—and we need your help. The question is this: Will our collective efforts—yours, mine, and the world's—work? It is for this very reason that this book has been written.

Support Vector Machines (SVMs)

By the 1990s, neural networks and machine learning were developing more and more sophisticated algorithms. Advances in machine learning, particularly in supervised learning and statistical methods, marked the 1990s. Support vector machines (SVMs) and decision trees gained popularity. SVMs are a class of supervised machine-learning algorithms that can be used for classification or regression tasks. SVMs are particularly well suited for tasks where the goal is to separate data points into different classes or groups. They were introduced by Vladimir Vapnik and his colleagues and involve several key components such as linear separation (i.e., separating data into different classes into "hyperplanes" to better identify the margins between two classes). The margins compose the distance between the hyperplane and the nearest data point from either class. The SVM then aims to maximize this margin for classification. Support vectors are the data points that are closest to the hyperplane and play a crucial role in defining the optimal hyperplane. These are the instances that, if removed or changed, could alter the position of the hyperplane. SVMs have been widely used in various applications, including image classification, text categorization, and bioinformatics. Their ability to handle both linear and nonlinear relationships, along with an emphasis on maximizing margins, makes them a powerful tool in the machine-learning toolkit.

Here's how an SVM would work for a binary classification task using a hypothetical scenario. The process involves training the SVM on historical data and then making predictions on new, unseen data. Here's a step-by-step explanation:

Task: Binary classification to predict whether a given e-mail is spam (positive class) or not spam (negative class).

Steps

1. **Data Collection**

   Collect a dataset of labeled e-mails where each e-mail is labeled as spam or not spam based on certain features (e.g., word frequency, presence of certain keywords).

2. **Data Preprocessing**

   Preprocess the data, extracting relevant features from the e-mails and encoding categorical variables. Split the dataset into training and testing sets.

3. **Training the SVM**

   Use the training set to train the SVM. During training, the SVM learns the optimal hyperplane that separates the spam and non-spam e-mails, maximizing the margin between the two classes.

4. **Feature Extraction for Real-Time Data**

   For real-time data, apply the same feature extraction process (e.g., term frequency–inverse document frequency [TF-IDF], which retrieves information by quantifying the importance or relevance of a string of words or phrases within a document) used during training to convert the incoming e-mail features into a format suitable for the SVM.

5. **Prediction in Real Time**

   Use the trained SVM model to predict whether the new e-mail is spam or not. The prediction variable now contains the SVM's prediction for the new e-mail (1 for spam, 0 for not spam).

Through marginal comparisons and categorizations, the SVM has become an extremely powerful and productive tool for machine learning (ML).

Decision Trees

Decision trees are a popular machine-learning algorithm used for both classification and regression tasks in artificial intelligence. Decision trees represent a flowchart-like structure in which each internal node represents a decision based on a particular feature; each branch represents the outcome of the decision, and each leaf node represents the final prediction or classification. These structures make it easy to visualize and understand how decisions are made within the model.

Figure 1.5 is an example of a decision tree (from Wikicommons) pondering the question of whether or not to upload an image from the internet.

You can see how the process works not only within ML systems, but also in our own decision-making situations.

A New Millenium: From Y2K to Armageddon

At the same time software development was booming in AI, the field of robotics was also making significant progress with the emergence of autonomous robots and computer vision systems. By the late 1990s and early 2000s, AI technologies began to be integrated into various applications, including web search engines, recommendation systems, and speech

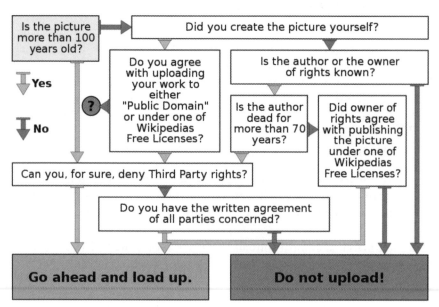

Figure 1.5.   Decision tree. *Wikimedia Commons.*

Figure 1.6.   Garry Kasparov. *Wikimedia Commons.*

recognition. Many readers may recall how, in 1997, IBM's Deep Blue defeated chess world champion Garry Kasparov, showcasing AI's capabilities in strategic thinking. The look of frustration on Kasparov's face when he realizes he has lost to a machine is timeless (figure 1.6).*

Remember that look; it's a hint at how we all will be looking if we lose the capacity to control the rapidly increasing intelligence of AGI.

The 2000s saw increased applications of machine learning in diverse fields such as finance, health care, and natural language processing. It was also a decade that focused much more on the potential for harm that AI might bring to the world. The Singularity Institute, established in 2000, became a pivotal organization studying AI risks and opportunities.

In 2000, Eliezer Yudkowsky founded the Singularity Institute for Artificial Intelligence with funding from Brian and Sabine Atkins, with the purpose of accelerating the development of artificial intelligence (AI). However, Yudkowsky began to be concerned that AI systems developed in the future could become superintelligent and pose risks to humanity, and in 2005 the institute moved to Silicon Valley and began to focus on ways to

---

* Although this picture is from a different match, it expresses very well the frustration we all might face if we are outsmarted and outplayed by a superintelligent AI.

identify and manage those risks, which were at the time largely ignored by scientists in the field.[71]

In late 2012, the institute sold its name and web domain to Singularity University—a nonaccredited university founded by Peter Diamandis and Ray Kurzweil—and, in the following month, was relaunched under the name Machine Intelligence Research Institute. Over the years, AI risk discussions have become more mainstream, reaching a broader audience of scientists and the general population, as evidenced by peer-reviewed journals focusing on the theme. This period saw the emergence of other prominent writers on AI risk, such as Ben Goertzel, who, like Yudkowsky, has made some rather depressingly stark and foreboding predictions regarding AI time lines and consequences. Goertzel heads up SingularityNET, which, according to its website,

> was founded . . . with the mission of creating a decentralized, democratic, inclusive and beneficial artificial general intelligence. An "AGI" that is not dependent on any central entity, that is open for anyone and not restricted to the narrow goals of a single corporation or even a single country. SingularityNET team includes seasoned engineers, scientists, researchers, entrepreneurs, and marketers. Our core platform and AI teams are further complemented by specialized teams devoted to application areas such as finance, robotics, biomedical AI, media, arts and entertainment.[72]

Both Yudkowsky and Goertzel began important discussions that expanded into the realm of philosophy, particularly regarding the implementation of moral values in artificial agents, leading to the evolution of fields like artificial morality and computational ethics. It is interesting to watch a TED Talk by Yudkowsky to hear him lament that he has been warning the public of the impending dangers of AGI—a term, by the way, that Goertzel's colleague, Mark Gubrud, invented.[73] While significant progress has been made in the last two decades, many challenges remain to be addressed in the pursuit of safe and beneficial artificial intelligence.

By 2010, Big Data and deep learning advancements in hardware and the availability of massive datasets led to the rise of deep learning (DL) and further developments in neural networks, enabling breakthroughs in computer vision, natural language processing, and speech recognition. Such networks allowed for DL, a subset of ML, to function more efficiently.

But what is DL?

Deep learning is a subset of machine learning, which is essentially a neural network with three or more layers. These neural networks attempt to simulate the behavior of the human brain—albeit far from matching its ability—allowing it to "learn" from large amounts of data. While a neural network with a single layer can still make approximate predictions, additional hidden layers can help to optimize and refine for accuracy. Deep learning drives many artificial intelligence (AI) applications and services that improve automation, performing analytical and physical tasks without human intervention. Deep learning technology lies behind everyday products and services (such as digital assistants, voice-enabled TV remotes, and credit card fraud detection) as well as emerging technologies (such as self-driving cars).[74]

As shown in figure 1.7, we can see the history of AI, leading from its definition to a further understanding of its various technical components such as ML, DL, and, finally, to generative AI.

In 2011, IBM's Watson showcased the capability of AI in natural language processing by defeating human champions—like Ken Jennings—in the game show *Jeopardy!* But it didn't just defeat the *Jeopardy!* champs; it

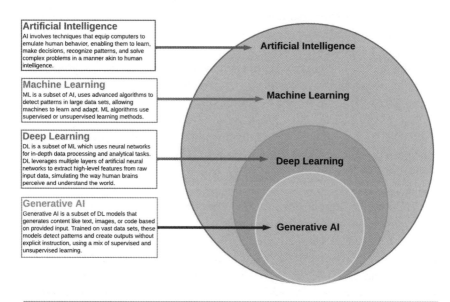

Artificial Intelligence
AI involves techniques that equip computers to emulate human behavior, enabling them to learn, make decisions, recognize patterns, and solve complex problems in a manner akin to human intelligence.

Machine Learning
ML is a subset of AI, uses advanced algorithms to detect patterns in large data sets, allowing machines to learn and adapt. ML algorithms use supervised or unsupervised learning methods.

Deep Learning
DL is a subset of ML which uses neural networks for in-depth data processing and analytical tasks. DL leverages multiple layers of artificial neural networks to extract high-level features from raw input data, simulating the way human brains perceive and understand the world.

Generative AI
Generative AI is a subset of DL models that generates content like text, images, or code based on provided input. Trained on vast data sets, these models detect patterns and create outputs without explicit instruction, using a mix of supervised and unsupervised learning.

Artificial Intelligence

Machine Learning

Deep Learning

Generative AI

Unraveling AI Complexity - A Comparative View of AI, Machine Learning, Deep Learning, and Generative AI.

(Created by Dr. Lily Popova Zhuhadar, 07, 29, 2023)

Figure 1.7.   AI to generative AI. *Wikimedia Commons.*

crushed them by scoring more than triple the closest competitors' score. And that was in 2011.

For the rest of the decade, various people and organizations made improvements in both computer capabilities and robotic evolution. Alex Graves made significant contributions to the field of sequence generation with his work "Generating Sequences with Recurrent Neural Networks" in 2013. Sequence generation in AI refers to the task of generating a sequence of outputs, often in a sequential manner, based on input data or context. This task is prevalent in various applications, including natural language processing, time series prediction, and music generation. The goal is to produce a meaningful and coherent sequence that follows a certain pattern or structure.

Meanwhile, Diederik P. Kingma and Max Welling proposed the variational autoencoder (VAE) in the same year, providing an important probabilistic framework for generating complex data. A VAE is a type of generative model in the field of artificial intelligence and machine learning. It is designed to learn a probabilistic representation of input data, allowing for the generation of new data points that resemble the training data. VAEs belong to the broader category of autoencoders, but they have specific characteristics that set them apart, including their ability to generate new samples through latent space interpolation. Variational autoencoders have proven to be powerful tools for generative modeling and unsupervised learning tasks. They have been applied in diverse fields, ranging from image synthesis and style transfer to drug discovery and anomaly detection.

In 2014, Ian Goodfellow and his colleagues Jean Pouget-Abadie, Mehdi Mirza, Bing Xu, David Warde-Farley, Sherjil Ozair, Aaron Courville, and Yoshua Bengio introduced the groundbreaking generative adversarial network (GAN) architecture. GANs consist of a generator and a discriminator network engaged in a competitive process to generate realistic data.

In 2015, research institutions like OpenAI and DeepMind have played pivotal roles in advancing generative AI for natural language processing (NLP). OpenAI developed models like "GPT" (generative pre-trained transformer)—culminating in GPT-3 in 2020, GPT-3.5 in 2022, GPT-4 in 2023, and GPT-4o in 2024—which showcased exceptional language generation capabilities. As noted earlier in this chapter, transformer technology possesses a type of neural network architecture, which uses an autoregressive language model that optimizes deep learning capabilities to produce human-like text. It was trained using forty-five terabytes of text data, including almost the entire public web.

In 2016, DeepMind's "WaveNet" introduced an impressive speech synthesis model based on deep neural networks. As we are currently witnessing, generative AI is a rapidly evolving field, and numerous researchers and scientists around the world have contributed to its progress. These include individuals from academia, industry, and open-source communities, who are collectively driving advancements in areas like style transfer, image synthesis, music composition, and more. Also in 2016, Google's AlphaGo defeated a world champion Go player, Lee Sedol, four games to one, highlighting the effectiveness of reinforcement learning in mastering complex games. This has been considered quite an extraordinary feat because of the level of complexity involved in playing the game. On their website, under the heading "Making History," the creators of this technology cite the following:

> Our artificial intelligence (AI) system, AlphaGo, learned to master the ancient Chinese game of Go—a profoundly complex board game of strategy, creativity, and ingenuity. AlphaGo defeated a human Go world champion a decade before experts thought possible, inspired players around the world to discover new approaches, and arguably, became the strongest Go player in history. It proved that AI systems can learn how to solve the most challenging problems in highly complex domains.[75]

Figure 1.8.   The game of Go. *Wikimedia Commons.*

For those familiar with the game of Go (figure 1.8), you will know just how complex it is.

For those who are not, here's a ChatGPT response when asked to explain it in three sentences:

> Go, an ancient strategy board game, is played on a gridded board with black and white stones. The goal is to dominate territory through strategic stone placement, capturing opponents' stones, and creating cohesive groups. Despite its straightforward rules, Go is celebrated for its profound strategic intricacies and boasts a rich history spanning over 2,500 years.*

The most important thing to keep in mind about the game is that it has an enormous number of possible moves. How many? This many:

> Given that a typical chess game has a branching factor of about 35 and lasts 80 moves, the number of possible moves is vast, about 35^{80} (or 10^{123}), aka the "Shannon number" after the Bell Laboratory pioneer Claude Shannon who not only invented information theory but also wrote the first paper on how to program a machine to play chess, back in 1950. Shannon's number, at 10^{123}, is huge, in particular considering there are only about 10^{80} atoms in the entire observable universe of galaxies, stars, planets, dogs, trees and people. But Go's complexity is bigger, much bigger. With its breadth of 250 possible moves each turn (Go is played on a 19 by 19 board compared to the much smaller eight by eight chess field) and a typical game depth of 150 moves, there are about 250^{150}, or 10^{360} possible moves. This is a number beyond imagination and renders any thought of exhaustively evaluating all possible moves utterly and completely unrealistic.[76]

So the sheer complexity of the game adds to its difficulty to master. And yet, AlphaGo was able to do this quite well—a feat many predicted would not be possible by AI. The Korea Baduk Association was so impressed that the group awarded AlphaGo the highest Go grandmaster rank, which was an honorary 9 dan (the highest ranking possible).[77] AlphaGo's success was attributed to the use of deep learning (DL), a specific type of artificial neural network (ANN). Today, artificial neural networks and DL serve as the fundamental building blocks for various AI applications, such as image recognition algorithms used by platforms like Facebook and speech recognition algorithms powering smart speakers and self-driving cars.

From 2017 to 2024, there were major advancements in all areas of AI. In hardware, there have been increasing advancements in graphics

---

* ChatGPT 3.5 search prompt: Explain the game of Go in three sentences.

processing units (or GPUs), which have been tailored for deep learning tasks. Google introduced its tensor processing units (TPUs), which are specialized hardware for accelerating machine-learning workloads, offering high performance and efficiency. In software, DL frameworks such as TensorFlow and PyTorch continued to dominate, with regular updates and improvements. Both frameworks saw increased adoption and community support. Transfer learning gained popularity, enabling the use of pre-trained models for new tasks, saving computational resources and training time.

There were also increases in computational power for training large-scale models, which has led to the use of powerful computing clusters and cloud-based solutions. While not yet mainstream, there were ongoing developments in quantum computing, with some potential applications for solving complex AI problems. Businesses have been using robotic process automation (RPA) and vastly improved integration of AI into various business processes to automate routine, rule-based tasks. Computer microprocessing or semiconductor chip development has improved and now includes neuromorphic chips, which mimic the structure and function of the human brain, for more efficient and specialized AI processing.

Several companies, including NVIDIA and Intel, invested in designing chips specifically tailored for AI workloads, optimizing performance for deep learning tasks. In late March 2024, the leather-jacketed CEO of NVIDIA—Jensen Huang[78]—revealed to a packed crowd at San Jose's SAP Center their latest and greatest creation: "NVIDIA's GB200 Superchip and its 'Blackwell' architecture. [The new chip] will offer up to a 30x performance increase compared to the NVIDIA H100 GPU for large language model inference workloads while using up to 25x less energy, which would make AI more efficient and cheaper."[79]

Huang is an advocate for rapid change in the AI game. He believed that general-purpose and accelerated computing was slowing down. With NVIDIA's new design, he believes the world is "seeing the start of a new industrial revolution."[80] Huang believes the industry needs to be more creative in finding ways to scale up ability *and* lower costs thereby allowing greater, more sustainable computing power to be available to everyone.

The development of this latest semiconductor superchip, which, by the way, contains 208 billion transistors and can be used for trillion-parameter scale generative AI, marks a unique point in the history of AI. As we are starting to notice a trend, many in the AI biz believe that scaling up computing power along with access to more and more data are the necessary and sufficient steps for building a superintelligent machine god (SMG).

This new NVIDIA chip will allow for a considerable increase in computing power. By the way, if we were living in a movie right now, this would be the point at which somebody from the future would come back to destroy the GB200 superchip.

Something very important to keep in mind is that there are only a few places around the world that manufacture semiconductor or microprocessing chips. At the top of the heap, there's Taiwan Semiconductor Manufacturing Co. (TSMC), followed by Samsung in South Korea, and Intel in the United States. The United States has quite a few companies now producing these valuable and essential products, with China rapidly advancing. You may remember how the COVID-19 global pandemic led to a supply-side shortage of semiconductors, which in turn led to higher demands for products dependent upon such valuable commodities (e.g., cars, computers, cell phones, etc.).

Meanwhile, in the field of robotics, there was steady progress in applying reinforcement learning to robotic systems, allowing robots to learn and adapt to their environments. It is truly amazing to see how clunky and clumsy the early versions of robots were compared to now. From the earliest days (where robots could barely take a few steps) to now (where they are doing backflips off tables and dancing as a choreographed group to "Do You Love Me?" by the Contours), we are seeing a very clear trajectory of the potentialities that await us in the near future. And of course, there have also been considerable developments in so-called weak AI, or ANI that came in the form of generative AI technology, and applications in relation to natural language processing (NLP), which showcased the ability to generate coherent and contextually relevant text. Let's take a look at some of the latest applications of weak ANI.

Practical Applications of Weak ANI Today

The most popular example of weak ANI today comes in the form of Transformers (figure 1.9).

No . . . not those transformers.

I'm talking about text, image, audio, and video generative transformers.

Text Transformers: Generative Pre-trained Transformers (or GPT for Short)

The type of transformers I'm referring to include ChatGPT, LLaMA, BERT, Claude, Bard, and many others. But what is meant by a "transformer" in this case? What do such things "transform"?

Figure 1.9.   Transformer robots. *Wikimedia Commons*.

Interestingly, when you ask an AI model to define GPT, here's one response:

> Generative pre-trained transformers (GPT) are a type of artificial intelligence (AI) model that can generate human-like text based on a given prompt or input. These models are trained on large datasets of text and use a self-supervised learning approach, which means they learn to predict the next word in a sequence of text without being explicitly taught to do so. This allows them to generate text that is coherent and contextually appropriate.[81]

Such forms of ANI incorporate large language models (LLMs), which use DL to scour the entire internet in an effort to understand, summarize, generate, and predict new content. They are a form of "generative transformer AI," which means they use neural networks similar to the human brain that are commonly referred to as transformers. With a large number of parameters and the transformer model, LLMs are able to understand and generate accurate responses rapidly, which makes the AI technology broadly applicable across many different domains.

But what are they used for? Here are just a few of the applications for which GPTs are used:

- Text Generation: GPT are advanced language models capable of autonomously creating text on a wide array of subjects based on their extensive training.
- Content Summary: GPT models can condense lengthy documents or articles into concise summaries, providing readers with a quick understanding of the main points.
- Translation: These models, when trained in multiple languages, exhibit proficiency in translating text from one language to another, facilitating cross-lingual communication.
- Sentiment Analysis: GPT models are equipped to assess the emotional tone of a piece of text, aiding in the interpretation of its underlying sentiment, whether it's positive, negative, or neutral.
- Rewriting Content: These GPT models possess the ability to rephrase or restructure text, enhancing its clarity or stylistic quality as needed.
- Classification and Categorization: They can classify and categorize text, helping to organize and group content by its subject matter or characteristics.

- Conversational AI and Chatbots: These models excel in natural language understanding and generation, making human–computer interactions through chatbots and conversational AI systems more seamless and human–like, setting them apart from earlier, less sophisticated AI technologies.

GPTs utilize three factors:

- Positional Encoding: The order of words in sentences.
- Attention: Learning grammar rules over and over.[82]
- Self-Attention: Disambiguate words, recognize parts of speech, and word tense.

Many of us have already experienced the wonders of this technology—from prompting it to write 1,500-word essays on Shakespeare's *Hamlet* in iambic pentameter, to prompting it to write code in a variety of programming languages, GPT technologies have quite literally transformed the way in which information is mined and exploited for a variety of uses. Numerous businesses are using such forms of ANI to help with various tasks and objectives.

Image Transformers

In much the same way that a GPT can create various original operations in text and code, programs like DALL-E, Midjourney, and Stable Diffusion create original images from textual descriptions. They use a twelve-billion parameter version of the GPT-4 transformer model to interpret natural language inputs. These programs use a convolutional neural network because it mimics the way the human brain processes vision. In other words, different layers of neural nets search out different aspects and patterns of a visual field through filters so that it can combine them all into a comprehensive understanding of what it is "looking at."* So, for example, you can prompt this system to create an image in the style of realist painters of a raccoon riding a bicycle next to the Grand Canyon, while eating an ice cream cone, and it will produce dozens of images (figure 1.10). Or, instead of realist painters, you could prompt it in the style of abstract expressionism, or surrealism, or any other artistic style you could possibly imagine.

---

* For a more comprehensive understanding of convolutional neural networks, see what ChatGPT 3.5 says about it in appendix B.

Figure 1.10.   DALL-E creation of raccoon on bike. *Created for the author. All rights to the author.*

Video Transformers

Perhaps the most powerful or visceral effects of this new technology can be found in its use to represent fake video images. The viral sensation that is the Tom Cruise deepfake[83] has left many wondering how long it will be before we can simply devise plots, characters, and actors and create original movies on demand. This new technology also has many of us wondering how it might be used to spread misinformation or, worse, disinformation—the intentional spreading of false information—against others in a number of ways. These represent just a few of the applications of weak ANI. We can only imagine what the future will bring. But they do paint a fairly clear picture that the technology is only going to improve and do so at an ever-increasing rate.

Today, AI-driven applications, such as virtual assistants and autonomous vehicles, have been gaining widespread attention and adoption. AI continues to advance rapidly, with research focusing on explainable AI, reinforcement learning, generative models, and AI ethics. AI is increasingly integrated into various aspects of daily life, from smart homes to health care, finance, and transportation. As noted earlier, AI has experienced periods of optimism (AI summers) and periods of reduced funding and interest (AI winters). Today, we are definitely in a summer, which may be extended to the great advantage of humanity, as well as many other species. On the other hand, if we are unable to harness and control its reach, we will have knowingly created a powerful god that we will no longer be able to control and can only hope it somehow decides to keep us around. For now, we can stay optimistic and look forward to the great opportunities and discoveries that lie ahead as AI continues to evolve and develop and plays out its integral role in shaping the future of technology and society. At this point, let's take a look at some of the current benefits of AI and speculate on what is yet to come.

The Benefits of AI   2

T his chapter is divided into two sections and for good reason. In the first section, we will consider several of the most current and proposed benefits that have been and are being developed from artificial intelligent technologies. But it is important to understand that these benefits are the result of ANI, or artificial narrow intelligence. These technologies fall under the rubric of "weak" AI because they are fully under human control. In other words, your Roomba is not going to "go rogue" and decide to threaten or kill you, your family, or the pet dog because it quite literally cannot. Their capacity to act outside of their programs, algorithms, or commands is so narrow that it is impossible to deviate from them. In the second section of this chapter, we will get a little more speculative. We will consider the potential for benefits that may be gleaned from the emergence of AGI (artificial general intelligence) or even ASI (artificial super intelligence). In this section, we will ponder what it would be like to have a god before us that will respond to our every command and desire—especially, the command or desire not to harm or kill us. What would you want such a god to do? What would be the first thing you would ask it, and why? We may want to start thinking now about the types of questions to ask such a god—because we're already well on our way to building it.

Part I: The Benefits of ANI Technologies
The proposed and evolving potential benefits from developments in artificial narrow intelligence are impressive. In an interview at the end of 2023, Reinvent Futures founder Peter Leyden,[1] when asked about the

coming years of AI technological developments, said some futurists believe
that generative forms of ANI will be thoroughly enmeshed within and
throughout all aspects of life:

> The explosion of positive uses for generative AI that will proliferate
> throughout the year as millions of entrepreneurs apply their creativity in
> myriad directions. You [have to] remember that AI is a general purpose
> technology that can and ultimately will be applied to almost everything
> over time, in every industry, every field. What would not benefit from
> applying machines that can now think? The closest thing we have in recent
> memory is the arrival of the internet, and AI is way bigger than that. The
> 1990s saw an explosion of startups as entrepreneurs from all over the world
> poured into the San Francisco Bay Area with crazy ideas about what to do
> with that new capability of connectivity. I was there back in the day, and
> I can tell you today San Francisco is every bit as energized with the even
> larger capabilities of AI.[2]

There is so much potential and so many AI developments occurring in
real time that it is difficult to keep up. Here are just a few:

- We are already witnessing improvements to health care in which
  AI programs are assisting in early disease detection, personalized
  medicine, drug discovery, and robotic surgeries, as well as analyzing
  vast amounts of medical data to develop better treatment plans.
- AI-powered tools can enhance education systems by providing
  personalized learning experiences, adaptive tutoring, intelligent
  assessments, and improving educational outcomes for students.
- Many businesses have already adopted the use of AI technologies to
  optimize industries like manufacturing, transportation, logistics, and
  energy, reducing costs, minimizing errors, and improving overall
  productivity.
- AI is accelerating scientific research by analyzing complex datasets,
  predicting outcomes, and assisting scientists in areas such as
  genomics, climate modeling, and high-energy physics.
- AI-powered smarter personal and virtual assistants can handle tasks,
  provide personalized recommendations, manage schedules, and
  improve daily productivity and convenience.
- AI is already enhancing transportation systems with self-driving
  cars, traffic optimization, and intelligent route planning, reducing
  accidents, congestion, and travel time.

- In regards to sustainability, AI can help monitor and manage natural resources, predict environmental patterns, optimize energy consumption, and enable more sustainable practices.
- As well, AI can improve language translation, speech recognition, and natural language processing, facilitating cross-cultural communication and breaking language barriers.
- AI will also increase accessibility by developing technologies like speech recognition, computer vision, and advanced prosthetics, which will promote greater accessibility and independence for people with disabilities.

Because of the explosion currently happening in the development of AI technologies, it is difficult to imagine the full range of potential contributions and benefits they will bring to humanity. By the time this book is published, there will be advancements in AI technologies that will just be emerging, and this book may require a second, third, and more editions in the future to fully consider them. What we can say at this stage is that the future looks promising in many ways and that our lives will indeed be improved by them—and quicker than you might think. There has been considerable work in the area of forecasting or predicting what benefits AI technologies may bring to humankind. Since 2015, there has been the "One Hundred Year Study of Artificial Intelligence," which considers, every five years, what has been developed and what is predicted to occur with AI technologies. The first report appeared in 2016 and is described in the following way:

> The One Hundred Year Study is modeled on an earlier report informally known as the "AAAI Asilomar Study." During 2008–2009, the then president of the Association for the Advancement of Artificial Intelligence (AAAI), Eric Horvitz, assembled a group of AI experts from multiple institutions and areas of the field, along with scholars of cognitive science, philosophy, and law. Working in distributed subgroups, the participants addressed near-term AI developments, long-term possibilities, and legal and ethical concerns, and then came together in a three-day meeting at Asilomar to share and discuss their findings. A short written report on the intensive meeting discussions, amplified by the participants' subsequent discussions with other colleagues, generated widespread interest and debate in the field and beyond.[3]

What I found most interesting about this first report was its lack of concern for the potential for AI existential risk from AGI/ASI developments. In fact, there was a noticeable downplay to this potentiality:

> Contrary to the more fantastic predictions for AI in the popular press, the Study Panel found no cause for concern that AI is an imminent threat to humankind. No machines with self-sustaining long-term goals and intent have been developed, nor are they likely to be developed in the near future. Instead, increasingly useful applications of AI, with potentially profound positive impacts on our society and economy are likely to emerge between now and 2030, the period this report considers.[4]

"No cause for concern that AI is an imminent threat to humankind." This sounds like it's right out of an action movie where the protagonist, a Bruce Willis type, is trying to warn politicians about an imminent threat—and, well, they're just not listening. Quite the dramatic scene. Except, this isn't a movie. I am again reminded of Sam Harris's eloquent plea that even if we had fifty years to get ready for AGI/ASI, we should probably start doing something about it now. Fortunately, my team and the rest of the crew at Convergence Analysis—along with far too few other collaborators around the world—are specifically working on this potentiality.[5]

There has since been a second report called "Gathering Strength, Gathering Storms: The One Hundred Year Study on Artificial Intelligence (AI100) 2021 Study Panel Report." In this report, a study panel responds to a collection of twelve standing questions (SQs) and two workshop questions (WQs) posed by the AI100 Standing Committee. Although again there is no explicit mention of or concern for AI existential risk in this report, the Standing Committee does make tacit reference to it in two of their standing questions. The first question asks: "What are the prospects for more general artificial intelligence?" Their response has not changed much since their 2016 report:

> Recent years have seen substantial new research, especially in the machine learning community, in how to imbue machines with common sense abilities. This effort includes work on enabling machines to learn causal models and intuitive physics, describing our everyday experience of how objects move and interact, as well as to give them abilities for abstraction and analogy. AI systems still remain very far from human abilities in all these areas, and perhaps will never gain common sense or general intelligence without being more tightly coupled to the physical world. But grappling with these

issues helps us not only make progress in AI, but better understand our own often invisible human mechanisms of general intelligence.[6]

Proposing that AI systems are "very far from human abilities . . . and perhaps will never gain common sense or general intelligence" seems a bit naïve. How is this quantification of time (i.e., "very far from human abilities") calculated or determined? How have they measured this? It's not as though we can predict, exactly, when AGI could develop. But to suggest that it remains very far away, and that this is perhaps because we have yet to understand our own "invisible human mechanisms of general intelligence," appears more poetic than scientific.

The second question to which the panel considers AI risk is this: "What are the most pressing dangers of AI?" You might think that *the* most pressing danger of AI would be existential risk. But again, this is not mentioned in the report. Instead, concerns are focused on the usual short-term risks of bias/discrimination, overdependence on automated decision-making, disinformation, threats to democracy, and so on. All of these are very serious risks, and we will consider many of them in the next chapter. However, they are all secondary and meaningless if we don't ensure that any and all forms of existential risks are prevented.

AI has the potential to bring about significant benefits to humanity across various domains. While it's important to note that the realization of these benefits depends on responsible development, deployment, and regulation, there are many ways ANI has and will contribute to benefit human welfare. Here are just a few of the more impactful ways AI has and will positively affect our lives.

Health Care Advancements

AI is revolutionizing health care by improving diagnostics, drug discovery, and personalized medicine. It can analyze vast amounts of medical data, assist in early detection of diseases, and aid in developing more effective treatments. When we consider just a few of the possible ways in which AI technology will improve our lives, we are optimistically encouraged to see how these will unfold. Currently, such technologies have already greatly advanced scientific and medical research. One of the most striking ways in which this has been manifested is through a program known as AlphaFold, which was developed by Google's DeepMind. For generations, scientists have labored over the structure of proteins and how they "fold" in order to carry out their various functions.

Every protein is made up of a sequence of amino acids joined together. These amino acids interact to form shapes, these shapes fold up on a larger scale to become complex 3D structures, basically the building blocks for all life on our planet. There are an unimaginable number of ways for proteins to potentially fold themselves. So figuring out how they quickly fold themselves in a consistent way is very tricky. Currently, there are over 200 million known proteins, with many more found every year. Each one has a unique 3D shape that determines how it works and what it does. The latest version (AlphaFold2) can now predict the shape of a protein, at large scale in a matter of minutes, and down to the accuracy of atoms! This is a significant breakthrough and highlights the impact AI can have on science. This work has been published in *Nature*, and has gone on to win Breakthrough prizes in Life Sciences, Mathematics and Fundamental Physics.[7]

It would take months or even years to learn about such proteins and their characteristics. But today, thanks to such technologies as AlphaFold, scientists can speed up this process, which will accelerate our understanding of many of the functions and systems of the human body. This, in turn, will rapidly increase the success for understanding and treating many diseases—including cancer. Now, it is an incredibly complex process and one which is quite difficult to explain in simple terms. But basically, it works something like this:

Data Input: Initially, scientists input data into AlphaFold, which usually includes the amino acid sequence of a protein and experimental data like the distances between pairs of amino acids.

Learning Patterns: AlphaFold then uses a deep learning neural network, a type of AI that, as we saw in the previous chapter, mimics the human brain's ability to learn. And learn it does. It learns by scanning a huge amount of protein structure data available in its databases.

Prediction: The program then predicts what the 3-D structure of the protein will look like if it were folded based on the input data. Based on the huge amount of input data, it then creates a computer simulation of how the protein might fold in on itself to create a specific shape.

Refinement: But the process doesn't end there. The program continues to make further predictions by using the laws of physics and other principles that govern protein folding.

Output: The final output is a detailed 3-D model of the protein's structure, showing how it folds and the spatial arrangement of its atoms (figure 2.1).

The significance of AlphaFold lies in its ability to predict protein structures with remarkable accuracy. These proteins act as little nano-machines

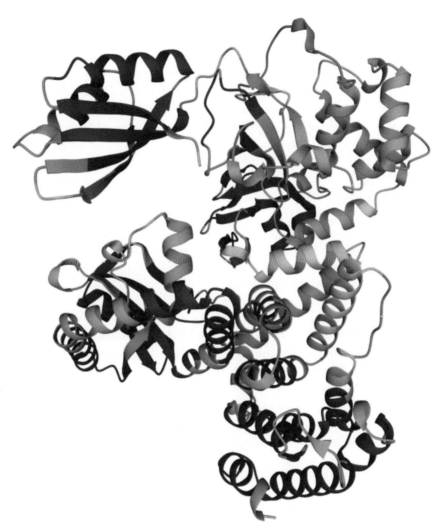

Figure 2.1.   Folding a protein. *Wikimedia Commons.*

that power every living thing on this planet. Demis Hassabis, CEO and founder of DeepMind, says that protein folding is one of the "Holy Grail-type problems in Biology."[8] Traditionally, determining a protein's structure was carried out experimentally through methods like X-ray crystallography or cryo-electron microscopy, which could be time consuming and challenging. AlphaFold accelerates this process, providing scientists with valuable insights into the functions of proteins and their roles in various biological processes.

A BRIEF HISTORY OF ALPHAFOLD

The challenge to come up with a breakthrough in this field has been going on since at least 1994 when something called CASP (Critical Assessment of Protein Structure) began competitions around the world. These competitions have been called the "Olympics of Protein Folding" and involved more than one hundred groups of dedicated teams of scientists from around the world. John Moult of the University of Maryland, the cofounder of CASP, stated: "When we started CASP in 1994, I certainly was naïve about how hard this was going to be." After a period of almost twenty-five years of trying to figure out how proteins folded, in April 2018, Moult said:

> In CASP 13, something very significant had happened. For the first time, we saw the effective application of artificial intelligence. The shapes were now approximately correct for many of the proteins, but the details, exactly where each atom sits, which is really what one would call a solution, were not yet there.[9]

By the time CASP 14 rolled around, it was spring 2020, and the world had been shut down due to the COVID-19 global pandemic. The Deep-Mind team began working on the SARS-CoV-2 protein called ORF8 at a distance, so to speak. They were, to say it mildly, extremely successful. As John Moult recounts:

> Here, what we saw in CASP 14, was a group delivering atomic accuracy off the bat—essentially solving what in our world is two problems: 1. How do you look to find the right solution? And then: 2. How do you recognize you've got the right solution when you're there? You know, these results were, for me, having worked on this problem so long after so many, many, stops and starts, and will this ever get there? Suddenly, this is a solution. We've solved the problem. This gives you such excitement about the way science works. About how you can never see exactly, or even approximately, what's going to happen next. There are always these surprises. And that really, as a scientist, is what keeps you going. What's going to be the next surprise?[10]

Ah, yes: "What's going to happen next?" Where have we read that before? We referred to this curiosity as a driving force of scientists and leaders of industry a bit earlier in chapter 1. It has driven humankind to discover some amazing things for the betterment and harm of ourselves and of our planet. Let us hope that our greatest capacity for learning and inquiry provides us with the greatest that AI has to offer while limiting or controlling the very worst. On July 28, 2022, DeepMind announced that AlphaFold had predicted the structures for nearly all proteins known to science (98.5 percent). And now, DeepMind has offered its findings to all scientists and researchers throughout the world:

> Last year, DeepMind released the source code of AlphaFold and made the structures of 1 million proteins, including nearly every protein in the human body, available in its AlphaFold Protein Structure Database. The database was built together with the European Molecular Biology Laboratory, an international public research institute that already hosts a large database of protein information. The latest data release gives the database a massive boost. The update includes structures for "plants, bacteria, animals, and many, many other organisms, opening up huge opportunities for AlphaFold to have impact on important issues such as sustainability, fuel, food insecurity, and neglected diseases," Demis Hassabis, DeepMind's founder and CEO, told reporters on a call this week.[11]

This advancement in protein folding by DeepMind is particularly crucial in fields like medicine, as understanding protein structures aids in drug discovery and the development of treatments for diseases. AlphaFold has already developed a more effective malaria vaccine, discovered new disease threats in Madagascar, developed new drugs to treat cancer, and is focusing its concerns to tackle antibiotic resistance.* AlphaFold represents a powerful tool that revolutionizes our approach to studying the very molecular machinery of life.[12]

MEDICAL DIAGNOSTICS

There have been and will continue to be other great developments in the fields of medicine through the advancements of AI, and there are many examples. AI can provide assistance in medical diagnostics by analyzing complex datasets such as medical images, genetic information, and patient records. This can lead to earlier and more accurate disease detection and personalized treatment plans.[13] AI algorithms can also accelerate the drug discovery process by analyzing vast datasets, predicting potential drug candidates, and optimizing clinical trial designs. And AI can also handle pedantic bureaucratic labor-intensive activities, which could free up the time doctors could be spending with their patients. Isaac Kohane—editor in chief of the *New England Journal of Medicine* and chair of Harvard's Department of Biomedical Informatics—stated:

---

* I don't think it's too much of a stretch to predict that such a development is definitely going to be a front-runner for the Nobel Prize in Medicine.

I think that doctors are not spending enough time with patients. They're spending too much time as bureaucrats. Having AI take care of that bureaucracy will allow doctors, we hope, to interface more with patients. That's a short term hope.[14]

It's interesting to see how Kohane echoes the very sentiments of what I had been trying to accomplish with the OSTOK Project[15] (i.e., the potential to take information about a patient and make medical inferences and treatments based specifically on their particular health conditions). In late May 2023, Dr. Kohane stated:

The longer term hope, which is when I say long term, I don't mean 20 years, I'm talking about five years, is that these programs will actually be able to look at all our data under the right privacy provisions and actually come up with new biomedical insights, new potential treatments, groups of patients who could benefit from these treatments and actually accelerate the drug discovery process as well. Because again, the same limitations that we're talking about human beings as doctors also afflict human beings as life science researchers, they can't know everything. They can't know of all the discoveries that are being made at one time. These programs are pretty good about knowing about everything.[16]

We will consider the speculative potential for AI health care assistance in the second part of this chapter. For now, what we can say with considerable confidence is that the world of health care is about to get a lot better in terms of diagnostics and treatments over the next few years.

Some of these scientific breakthroughs are occurring in real time right now. For example, Stanford computer science professor Daphne Koller believes that AI is currently revolutionizing the way in which various forms of cancer are diagnosed. Koller is also an AI pioneer and cofounder and CEO of the life sciences AI firm Insitro.[17] When asked what her research involves, Koller said:

"[W]hat we've taken on as an effort is to really learn the language of histopathology [the study of tissues] . . . and then use that to [. . .] give us potential [drug] targets." By using machine learning, the computer will "really learn the language of histopathology," she said, which in turn lets the machine predict genetic changes in patients with cancer with 90 percent to 95 percent accuracy. "So, basically, by looking at a slide, you can say this patient has this genetic mutation versus this other patient, something that no clinician can really do," she explained.[18]

The use of AI in biology and cancer research has created an entirely new field in science known as digital biology. This latest field is just one of many that will be developed as the history of science evolves:

> "Think about the late 1800s and chemistry, with the uncovering of the periodic table [of elements]," she offered, "Or the early 1900s, of course, physics, with the connection between energy and matter, space and time." The 1990s saw a similar explosion of discovery in two disciplines, said Koller, "Data/machine learning/AI, which is really something that began back then, and quantitative biology, which is the ability to measure biology at unprecedented fidelity." Those two disciplines are now merging, she said, to create a new field called digital biology, which is "the ability to read the biology digitally at this incredible fidelity at an unprecedented scale, interpret what we see using tools such as machine learning and AI, and then write biology using techniques like CRISPR and combinatorial chemistry, and all sorts of other things to make biology do things that it wouldn't otherwise do."[19]

Koller is confident and hopeful that AI will contribute not only to the medical fields of science but also to other fields, stating that we will see vast improvements in the environmental sciences, energy, bio-materials, and sustainable agriculture, as well as many other disciplines that will help make our world a better place.[20]

One recently developed cancer diagnostic tool is called PANDA (pancreatic cancer detection with artificial intelligence). With a high degree of accuracy, this new form of AI technology is able to identify and classify pancreatic lesions for the early detection and treatment of pancreatic ductal adenocarcinoma (PDAC). It is extremely important to detect pancreatic cancer as early as possible to begin treatment. Since most cases are discovered too late when the cancer has progressed into its third or fourth stage, early detection allows treatment to begin sooner, which is prolonging human lives considerably.[21]

> Pancreatic ductal adenocarcinoma is the most malignant form of solid carcinoma, with a mortality rate of over 450,000 each year. The high mortality rate, however, is largely because PDAC is often detected in the late stages when it is inoperable.
>
> Cases where PDAC is detected incidentally or early have a better prognosis and early treatment often results in substantial improvements in the survival rates of patients. The median overall survival rate in cases where PDAC has been detected and treated in the early stages is 9.8 years compared to the 1.5-year survival rate for most late-detection cases.[22]

This is great news for researchers, physicians, and patients alike. Speeding up the detection of early cancer development is crucial in stopping its spread through advanced treatments.

Another amazing advancement in AI medical technology for cancer research is IBM Watson for Oncology. Basically, it's an AI system that analyzes large volumes of medical literature and clinical trial data to provide personalized treatment recommendations for cancer patients. It assists oncologists in making informed decisions about the most effective treatment options. From the IBM Watson for Oncology Sales Manual we find the following definition of its services:

> IBM Watson for Oncology is software as a service (SaaS) that delivers an advanced ability to analyze the meaning and context of structured and unstructured data in clinical notes and reports, easily assimilating key patient information written in plain English. By combining attributes from the patient's file with clinical expertise from Memorial Sloan Kettering, external research, and data, Watson for Oncology identifies and ranks potential treatment plans and options. . . . Its latest capabilities include expanded coverage to include breast, lung, colon, rectal, gastric and cervical cancer.[23]

Just as I had wanted to build a similar machine (the OSTOK Project) to aid in medical diagnoses and treatments by bringing together information from a variety of patient-related sources, which echoed the decades-earlier work of Edward Shortliffe's MYCIN program, so too are we now seeing this idea—and hope for a better medical system—come to greater fruition. We have been making a continual move toward greater and greater personalized medical programs for patients. That is, all humans share similar biological and medical qualities and characteristics. However, how each one of us responds to various chemical compounds, treatments, and interventions is quite specific. Learning more about a patient's specific needs through cross-referencing AI technology speeds up the process and greatly facilitates any physician's ability to help their patient.

In terms of speeding up diagnostic abilities, Digital Diagnostics is using AI to identify eye problems more rapidly in those suffering from diabetes. In referencing their latest machine—the IDx-DR—it has been described as "a Class IIa Medical Device and the first ever FDA De Novo cleared autonomous AI."[24] The company's motto for such a device is "Close care gaps, prevent blindness." With the development of such an accurate and helpful diagnostic machine, less time is wasted in other stages of eye examination. The company further states that "IDx-DR is an AI diagnostic system that autonomously diagnoses patients for diabetic retinopathy (including macular edema)," which is a leading cause of blindness in diabetic patients. Essentially, the system analyzes retinal images to identify signs of the condition, allowing for early intervention.

And then there's PathAI for Pathology, an AI platform designed to assist pathologists in diagnosing diseases from pathology slides. It helps improve accuracy and efficiency in identifying diseases such as cancer by analyzing and interpreting medical images.

And yet another example of how AI is improving lives in health care advancements comes from just outside my hometown in our neighboring city: Cambridge, Ontario. Cambridge Memorial Hospital (CMH) became the first Ontario hospital to use artificial intelligence (AI) screening technology for clinical use in the assessment of low bone mineral density (BMD). The chief of diagnostic imaging at CMH, Dr. Winnie Lee, says this new AI diagnostic technology will help her assess BMD in patients more efficiently and in greater detail than ever before: "It's a silent disease—osteoporosis is under-recognized. According to Osteoporosis Canada, 20 percent to 25 percent of people don't realize they're at risk."[25] This new diagnostic tool, called Rho, was developed by radiologist Dr. Mark Cicero and his team at 16 Bit—a tech company whose mission is "to create trusted AI Software-as-a-medical device to improve the quality, efficiency, and accessibility of healthcare for all people."[26] Noticing shortfalls among industry practices in radiology, Cicero and his team found that "osteoporosis screening programs were underutilized and many patients would end up in the emergency department with a fracture who had never been screened before, we realized that something had to be done about this epidemic. [Rho] can automatically analyze X-rays done in patients over the age of 50 for any clinical indication to assess their risk of having osteoporosis or low bone density."[27] We are about to see an explosion in diagnostic applications that use AI. Humans will still be necessary to check the results of the AI. But the diagnostic forms of AI being developed now are simply faster, more thorough, less costly, and a thousand times more accurate than humans. The world of medicine is changing rapidly in real time, right before our eyes.

NEW DRUG DISCOVERIES
And AI will make our lives better by discovering better drugs. Insilico Medicine is one such company aiming to do this. They are using AI to accelerate the drug discovery process by employing generative models to design novel drug compounds and predict their potential efficacy. This,

in turn, leads to more efficient drug development. Their mission statement maintains that they want to "extend healthy productive longevity by transforming drug discovery and development with generative artificial intelligence, significantly reducing the time and cost to bring life-saving medications to patients."[28] These are some fine and noble ideals to pursue. And they are the type of virtues that truly gives us hope for the great accomplishments AI will bring to the field of medicine. But we might ask this question: How, precisely, do they propose to accomplish this goal?

The company has managed to coordinate a multidisciplinary platform that will accelerate the process of medical research and discovery. Keep in mind that this is just one company out of dozens, if not hundreds, of companies all quickly adopting AI systems to the way business is done. Their common bottom lines may simply be $$$, but if they are able to reduce human and animal suffering in the process, more power to them.

AI-ASSISTED SURGERY
The final way in which AI is benefiting health care that we'll consider is through precision robotic surgery. There have been several extremely impressive advancements in how surgeries are conducted today through the incorporation of AI:

> AI-assisted robotic surgery is an emerging field that holds great promise in improving surgical precision and outcomes. Surgeons can use robotic systems equipped with AI algorithms to perform complex procedures with enhanced precision, reduced invasiveness, and improved post-operative recovery times. AI can assist in real-time image analysis, providing surgeons with augmented information and guidance during procedures, thereby reducing the risk of complications.[29]

One such AI-assisted robotic system is the da Vinci surgical system, which allows surgeons to control robotic arms deft enough to peel and sew a grape with impressive accuracy and speed. Another is the Versius System. The benefits of this type of system is that it offers minimal access surgery (MAS), which means no scalpel usage, and this translates to reduced pain, fewer surgical complications, and fewer surgical site infections. Surgeons can plan entire robotic procedures and control the robotic arms while seated ergonomically and away from the patient rather than being hunched over the patient for hours on end. One of the most noted compliments of AI-assisted robotic surgery developments comes from the surgeons themselves who believe such modifications have reduced their postoperative muscle strains—particularly, lower back, knees, shoulders, and wrists—considerably. Versius's website offers this description:

> Our suite of fully-wristed instruments, combined with enhanced 3D HD vision, give surgeons a high level of accuracy when performing complicated procedural steps or operating in hard-to-reach areas. With high levels of precision and control, surgeons are able to perform more of their complex procedures through a minimal access technique. With the Versius digital ecosystem, we give surgeons the tools to understand the full clinical story of their procedures, so they can make informed decisions to improve patient care.[30]

The Versius unit has been used extensively at Milton Keynes University Hospital (MKUH) in Eaglestone, England, where surgeons have been using this form of AI technology for several years now:

> MKUH were the first in the UK to install Versius in 2019. In 2022, they reported an annual saving of 450 bed days following the adoption of Versius. But what does that look like for the teams using Versius every day? The implementation of surgical robotics has had a significant impact on patient outcomes. Enabled by the robot, the surgical team have been able to offer a minimally access approach to a wider range of patients—with benefits including helping to get patients back at home and returning to normal activity faster. The small and modular design of Versius has also enabled the team to quickly roll-out a multi-specialty robotics programme so as many patients as possible can benefit—and have plans to extend to further specialties. The surgeons using Versius have also reported less pain compared to when they operate with traditional laparoscopy. Some even feel this has helped them avoid injury and will extend their surgical careers. A more efficient and cohesive team performance has been observed in the robotic theatres—and the hospital has recorded a positive impact on staff recruitment and retention.[31]

With the increased need to free up beds in hospitals, it is heartening to see how AI technologies are helping people get treated quicker, better, and with less time spent recovering in hospitals so they can get back on with their lives.

As more and more health care AI technologies continue to advance, it will become much easier for patients to get the necessary treatment they need years—or, perhaps, even decades—sooner than they would under our current existing screening technologies.

A FINAL THOUGHT ON AI MEDICINE

It has always been my professional belief that, just as humans built instruments like telescopes to help them see farther toward the edge of the universe and they built microscopes to see farther downward into infinitesimally smaller worlds, so too should we use tools and instruments that can think much faster and more accurately than we ever could. AI technology has begun to revolutionize the way medicine is done. Forget about self-driving cars for now; AI technology in medicine is going to make our lives a lot better—and it's already happening.

Language Translation and Communication

AI-powered language translation services can improve our lives by allowing us to better connect the world by breaking down language barriers, which, in turn, fosters better global communication and collaboration. This has implications for international business, diplomacy, and cultural exchange. At this point in time, there have been many predictions on how AI will disrupt specific areas in industry. Some occupations have been predicted to be largely replaced by generative AI (GenAI). One such occupation is that of a translator. But some within the field disagree. For example, Paul Carr, CEO of Welocalize, one of the world's largest language services providers, has maintained that the translation business has always been on an increasingly upward trajectory when it comes to using new forms of assistive technology: "The industry was an early adopter of automation tools, with machine learning algorithms designed specifically for translation—namely, neural machine translation (NMT)—being widely used."[32] Carr maintains that rather than disrupt and replace translators, GenAI will instead enhance the accuracy of the translators who, he believes, will still be needed to provide overview and insight into an ever-improving system: "In summary, the application of GenAI into current translation workflows, including conducting translation itself, is less likely to result in a tectonic shift in efficiency than an uptick in the relentless grind toward more automation, a pathway the industry has been on for many years."[33] In this way, Carr believes translators will monitor the great volumes of work by GenAI and be able to consider and monitor bias, factual accuracy, toxicity, adherence to privacy regulations, and protection of intellectual property rights.

> Linguists and translators, therefore, will likely see an acceleration from doing raw translation work to working alongside automation tools to validate, verify and post-edit multilingual content generated directly from NMT and LLM algorithms. Content creators and authors will need

translation professionals more than ever to handle the increase in content volumes and to navigate the assessment, review and quality control of LLM-generated content.[34]

And Carr is not alone in thinking that there will be a joint effort between humans and AI programs when it comes to translation: "the best strategy might be a mix of human and machine translation. While humans can concentrate on more complex and nuanced translations, machines can assist in automating the translation process and handling simpler translations."[35]

Still, there are current learning companies that are making claims that their programs are already accomplishing the nuanced distinctions that Carr and others believe to be missing in language translation. Although not alone in this regard, companies like Harbinger[36] make some rather promising claims that their AI systems can accomplish extremely human-like end results such as the following:

1. Customer-Specific Content Translation

   On e-learning platforms AI-powered translation prioritizes individual preferences, language backgrounds, and cultural nuances to create a more inclusive and engaging educational environment for customers.

2. Translation of e-Learning Courses

   AI-based automated translation streamlines the translation of e-learning courses by efficiently converting text, audio, video, and multimedia elements from one language to another, making educational resources accessible to the target audience, irrespective of linguistic diversity.

3. Automated Multilingual Translation

   AI-driven translation can automate the process of translating learning content into multiple languages and translates documents into different languages and formats like Docx, PPTx, XLSx, and PDF, without compromising document formatting, enabling quicker dissemination of knowledge across various regions and expanding learning materials' reach to a broader, diverse audience.

4. Translation for Localization and Globalization

AI-driven translation technology accurately and contextually trans-lates text, graphics, and multimedia elements to cater to the local language and cultural sensitivities, which rapidly extends their reach to global markets.

5. AI-Powered Audio and Video Translation

AI can seamlessly translate audio and video content in e-learning modules to ensure a seamless multimedia-based learning experi-ence, which includes AI-based audio-to-video translation and vice versa, automatic face-to-face video translation with lip movement synchronization, translation of text in static video images, and auto-mated voice-based video translation and dubbing.

6. Translation from Authoring Tools

AI-based automated translation technology can be integrated easily into popular authoring tools like Adobe Captivate, Articulate Story-line 360, and DominKnow One. This can help businesses generate e-learning courses and corporate training material in their native language and automatically translate them into desired languages, saving time and resources.

7. AI-Based Translation for Accessibility

AI-driven translation significantly facilitates sign language translation and can create PDF documents accessible to the blind population, allowing them to be read using specialized software.

This list represents many of the ways in which translation software from generative AI can improve upon already-existing systems and benefit others in understanding information in their language of choice. This will have a substantial impact on the effectiveness of communication throughout many facets of society. In echoing the sentiments of Carr and others, within the final "End Note" of Harbinger's claims they cite: "Additionally, collabora-tion between AI and human translators can enhance the quality of transla-tions and provide a more holistic approach to eLearning."[37] So, it would appear—for now, at least—that translators have some job security. The ques-tion to consider at this point is this: For how long? That is, will generative AI systems just continue on a trajectory of self-improvement until they not only match human translation abilities, but surpass them?* And if so, what then? It is to this question—that is, the idea of massive job losses due to AI—and many more, that we will turn our attention toward in the next chapter. For now, let's look at how AI might provide benefits in communication.

There are several ways that generative AI has already made headway into specific areas of communication. For instance, within the field of mental health, AI has made considerable breakthroughs in reaching those with autism. One example involves the use of ChatGPT as a conversational partner. Those who are autistic have found that conversing with ChatGPT is more productive than communicating with human partners.

> For some, it's a place to chat about their interests when other people grow bored, or to work up social scripts to help them navigate conflict. It's also a new resource to turn to for support. Unlike a therapist or social worker, the bot is always available and doesn't bill by the hour.[38]

Many in the autistic community have also found that discussions with ChatGPT made them feel more empowered about social interactions:

> Using a chatbot to help with communication may seem unconventional, but it's in line with some established ideas used in social work to help people become more independent. "We talk about empowering people and helping people to be fully autonomous and experience success on their own terms," says Lauri Goldkind, a professor in Fordham University's Graduate School of Social Service who focuses on the marriage of social work and technology. An accessible tool like a generative AI bot can often help bridge the gap left by intermittent access to mental health services like therapy, Goldkind says.[39]

At this point in time, it is important to understand the limitations of ChatGPT, of which most of us will already be quite familiar. ChatGPT "can produce biased, unpredictable, and often fabricated answers, and is built in part on personal information scraped without permission, raising privacy concerns."[40] Professor Goldkind advises that people who use such services become adept at using it and understanding its limitations. Others, such as Margaret Mitchell, suggest that

> people who face more complex issues or severe emotional distress should limit their use of chatbots. "It could lead down directions of discussion that

---

* For the record, the day after I wrote this section on the benefits of AI for translation, I received an e-mail in my Junk Folder advertising the following product: "Instant Two-Way Translator. Portable Real-time Translation in 36 Different Languages. Perfect as a Pocket Dictionary and for Learning, Travel and Business Communications."

are problematic or stimulate negative thinking," she says. "The fact that we don't have full control over what these systems can say is a big issue."[41]

Although many such bots advise its users to seek out professional, human help when faced with more pressing mental health issues such as suicidal thoughts, there is always going to be the risk of harm—even when none is intended. Such forms of AI technology generally, but not always, direct an individual to sources of professional support—like emergency hotlines—if they detect a more serious nature to an individual's conversation. Again, we must understand that we are still very much in the early stages of these forms of technology. As they continue to improve, so will our reliance upon them grow. We will speculate on some potential scenarios for the future toward the end of this chapter.

The idea of using AI to help with mental health issues has, as we saw in the first chapter, been around since the 1950s with Joseph Weizenbaum's program ELIZA. As you may recall, Weizenbaum's secretary was convinced that the program was helping her deal with her personal psychological issues. In a recent study,[42] researchers have found that later versions of ChatGPT give better personal advice than professional columnists because it offers options that make it appear to be more empathetic (i.e., as though it is really listening to an individual's concerns).

> Providing advice of a personal nature requires a certain level of empathy (or at least the impression of it). Research has shown a recipient who doesn't feel heard isn't as likely to accept advice given to them. They may even feel alienated or devalued. Put simply, advice without empathy is unlikely to be helpful. Moreover, there's often no right answer when it comes to personal dilemmas. Instead, the advisor needs to display sound judgment. In these cases it may be more important to be compassionate than to be "right."[43]

Piers Howe, the lead author of the study, found that ChatGPT 3.5 was not very empathetic, ethical, or displayed sound judgment and so performed poorly when providing social advice. By focusing purely on solving an individual's problems and not addressing their emotional needs, individuals felt as though they were not being heard. This changed with the introduction of ChatGPT-4:

> The latest version of ChatGPT, using GPT-4, allows users to request multiple responses to the same question, after which they can indicate which one they prefer. This feedback teaches the model how to produce

more socially appropriate responses—and has helped it appear more empathetic.[44]

Howe and his colleagues conducted the study by randomly selecting "50 social dilemma questions from ten well-known advice columns covering a range of topics including relationships, ethical dilemmas, management problems, end-of-life issues, and many everyday qualms we can all relate to." They then presented each of the 400 participants with an ethical dilemma. The subjects were then presented with responses from both ChatGPT and a professional advice columnist and blinded the study by not revealing where each response was coming from. "About three-quarters of the participants perceived ChatGPT's advice as being more balanced, complete, empathetic, helpful and better overall compared to the advice by the professional."[45] Based on these results, it would appear that ChatGPT has passed a form of the Turing Test. It has also demonstrated that it is an extremely effective communicator. It is important to remember this as we progress throughout the book. For, there are many within the AI fields who believe that an AGI or ASI would be so effective at communication that we would be overwhelmed and completely outwitted by its vast and unimaginably powerful abilities to communicate—perhaps, deceptively so; and perhaps in directions that will generate unwanted harm to ourselves, other species, or even the planet itself. So keep both the notions of translation and communication abilities in the back of your mind as we move onto the potential for misuse and abuse of such AI technologies in the following chapter. For now, let's consider what benefits might befall business and industry with the advancements of AI.

Business and Industry—Automation and Efficiency

There is little question that AI technologies have for decades and, in a variety of ways, already transformed the labor landscape. We just need to consider for a moment how robotics has optimized productivity within so many industries. From the automobile assembly lines to Amazon, from customer service bots to remote-controlled systems, and so much more, AI has been automating routine and mundane tasks, allowing humans to focus on more complex and creative aspects of their work for decades. We know how this type of automation can lead to increased productivity and efficiency across industries. But at what cost? Will this not put more humans out of work? There is no question that new AI technologies will disrupt many businesses; and negatively. And we will look at some of these negative side effects in the next chapter. But for right now, let's consider how many believe there will be just as much opportunity for individuals to adapt to the changing labor landscape.

For example, as AI continues on its current evolutionary trajectory, Ben Goertzel, CEO of decentralized artificial intelligence marketplace SingularityNET, believes that allowing AI to take care of the mundane chores and tasks that humans currently perform will free up our time to pursue other, more noble, objectives. Goertzel believes "there are far more rewarding things for human beings to do than [scrambling] around to get resources."[46] He envisions a time in the not-so-distant-future where humans will have ample free time to explore their own interests and live comfortably through a form of universal basic income (UBI) similar to that proposed by 2020 presidential candidate Andrew Yang.[47] Okay, so we may see a utopian future with AI taking care of our mundane tasks and us living carefree lives. But until we get there (if we do at all), how, specifically, is AI benefiting us in business and industry?

Some believe that AI is already helping businesses perform better in a number of key areas:

> The trends in business that we see today in AI are largely coming with the increase of efficiency and productivity for many firms. Through the distribution of AI to different open AI models, more companies are exploring the possibilities to have some form of machine learning in their operations. Many years ago, in trying to improve manufacturing and production, these improvements were related to the physical world. We now see a very similar phenomenon happening with service companies; they're trying to use AI to enhance what they do. For example, marketing today is much more about digital interfaces because technology made it possible. Now with AI as part of the equation, it changes even more—interfacing with customers, profiling, the kinds of campaigns, and the algorithmic governance that we have.[48]

There are many different areas within business and industry that have been affected by advancements in AI technology. The clearest example for improvement in businesses through AI incorporation comes with automation and efficiency. AI has been and will continue to enable businesses to automate routine tasks, and improve operational efficiency while reducing human error. This can occur through process automation in which AI technologies like robotic process automation (RPA) help reduce human manual labor with automated repetitive and rule-based tasks. Although this could easily allow employees to focus on more strategic and creative aspects of their work, currently there are emerging sophisticated forms of ANI that will be able to assist with some of the strategic and creative aspects of work in both business and industry. Seen in this light, ANI is more of a facilitator of ideas, and its input may very well extend into more human-centered avenues.

The automation of manual processes leads to increased speed and accuracy in task execution. This, in turn, could easily lead to workflow optimization whereby AI-driven workflow management systems can analyze and optimize business processes. Such AI-driven technologies can better identify bottlenecks, streamline workflows, and use the most appropriate resources for optimal outcomes, leading to an improvement in the overall operational efficiency within companies.

In industrial settings, AI technologies are used for the predictive maintenance of equipment and machinery. In other words, sensors and AI algorithms can predict when equipment is likely to fail, which allows for proactive maintenance and a reduction in downtime. As well, AI can enhance supply chain efficiency by optimizing inventory management, demand forecasting, and logistics. Predictive analytics and machine learning algorithms help businesses anticipate changes in demand and adjust their supply chain accordingly, which reduces cost, time, waste, and energy. AI also accelerates data processing and analysis, which enables businesses to extract valuable insights from large datasets. Machine-learning algorithms can then identify patterns and trends that may not be immediately apparent to human analysts and used to make more accurate predictions regarding all aspects of a company's functions. AI technologies can also facilitate the automation of document processing tasks, such as data extraction and categorization.

Optical character recognition (OCR) and natural language processing (NLP) are used to automate the handling of unstructured data to increase efficiencies across the board.

In the services industries, chatbots have become so sophisticated and nuanced that they are in many uses involving customer service automation. AI-powered chatbots and virtual assistants automate customer interactions, which enhances responsiveness, ensures 24/7 support, and provides instant responses to queries. AI is employed in automating financial and accounting tasks, including invoice processing, expense management, and reconciliation, which reduces errors, improves accuracy, and speeds up financial reporting.

Within human resources (HR) throughout the world, AI is used in automating aspects of hiring and training, such as resume screening, candidate matching, and employee onboarding. Automated HR processes, which have been properly screened for improper biasing, save time and resources while ensuring a more objective evaluation of candidates.

And finally, in industries with significant energy consumption, AI can help optimize energy usage by analyzing patterns and adjusting consumption in real time. Smart building systems use AI to regulate lighting, heating, and cooling for energy efficiency. These applications of AI in automation and efficiency contribute to cost savings, improved productivity, and a more agile response to dynamic business environments—not to mention that it's eco-friendly.

Education

Has AI evolved to the point where it can replace teachers? No. Not yet. Will AI evolve to the point where it will replace teachers? That's a possibility. I personally prefer a face-to-face live lecture as opposed to an online version—both giving and receiving. But maybe my preferences are simply quaint and the future of education will resemble little of what we have experienced in our lifetimes. We'll just have to wait and see. For now, let's consider how AI is being used in various educational settings.

PERSONALIZED LEARNING

AI programs can tailor educational content to personalized individual learning styles, helping students grasp concepts more effectively. AI algorithms can analyze students' learning patterns and adapt educational content to suit individual needs and preferences. Personalized learning platforms provide tailored exercises, assessments, and feedback, allowing students to learn at their own pace. At the 2023 Stanford University AI+Education Summit, participants, including Stanford researchers, students, and industry leaders, deliberated on the transformative potential of AI in enhancing education. The event was the joint effort of both the Stanford Accelerator for Learning and the Stanford Institute for Human-Centered AI. The discussions at the inaugural event not only highlighted the positive impacts AI could have on education but also addressed the accompanying risks. This forum served as a platform for stakeholders to explore the intersection of artificial intelligence and education, emphasizing both the opportunities for improvement and the need to navigate potential challenges in implementing AI technologies in educational settings. They were able to identify what have since become common areas in which AI can actively assist or enhance education.

Several themes were identified, the first of which was how AI might enhance personalized support for teachers at scale. In looking for ways in which teachers could receive feedback quicker and more effectively, the attendees found that AI can effectively support teachers in a number of ways:

- **Simulating Students:** AI language models can serve as practice students for new teachers. Percy Liang, director of the Stanford HAI Center for Research on Foundation Models, said that they are increasingly effective and are now capable of demonstrating confusion and asking adaptive follow-up questions.
- **Real-time Feedback and Suggestions:** Dora Demszky, assistant professor of education data science, highlighted the ability for AI to provide real-time feedback and suggestions to teachers (e.g., questions to ask the class), creating a bank of live advice based on expert pedagogy.
- **Post-teaching Feedback:** Demszky added that AI can produce post-lesson reports that summarize the classroom dynamics. Potential metrics include student speaking time or identification of the questions that triggered the most engagement. Research finds that when students talk more, learning is improved.
- **Refreshing Expertise:** Sal Khan, founder of online learning environment Khan Academy, suggested that AI could help teachers stay up to date with the latest advancements in their field. For example, a biology teacher would have AI update them on the latest breakthroughs in cancer research, or leverage AI to update their curriculum.

For anyone who has taught or knows someone who has, you know how difficult it is to be a good educator. Any help in this area—such as that produced by AI—will be welcome aid.

Another area the Stanford Summit attendees considered was the extent to which AI will be an extension of learning or a replacement to learning. The analogy of the calculator to mathematics was considered in both positive and negative ways.

Today, the calculator is ubiquitous in middle and high schools, enabling students to quickly solve complex computations, graph equations, and solve problems. However, it has not resulted in the removal of basic mathematical computation from the curriculum: Students still know how to do long division and calculate exponents without technological assistance. On the other hand, Reich noted, writing is a way of learning how to think. Could outsourcing much of that work to AI harm students' critical thinking development?[49]

Ah, critical thinking. Is this even being taught in schools? We like to think it is. But is it? And by whom? I have spent much of my entire academic career in an attempt to bring critical thinking into the high schools of Ontario, Canada.[50] Everyone—from students, to parents, to teachers, to principals, to directors, to ministers of education—believed it was a great idea. And our efforts were working. Right up until the next election when the new government took over and scrapped the project. And then COVID-19 hit. We will never cease our efforts to bring critical thinking into the high school curriculum and, hopefully, to the elementary and middle school levels soon after. It would be wonderful to see how AI might help facilitate this.

Noah Goodman, associate professor of psychology and computer science, has raised an interesting question about the comparison between AI and a calculator. He suggested that AI (e.g., ChatGPT) might be similar to the printing press, which, instead of replacing human writing skills, made knowledge accessible to more people, similar to how this new technology could make certain tasks more accessible without eliminating the need for human skills. For the record, almost every new form of technology has been met with some skepticism about how it would interfere with education. Socrates was not a huge fan of people writing stuff down. It should all be committed to memory, according to him. Then, some opposed the writing of books; then, the printing of books, and now, AI.

A fascinating psycho-social aspect of education was discussed at the summit, which included considering how an artificially intelligent evaluator might seem less "judgy" than a human one:

> [F]or most students, fear of judgment from their peers holds them back from fully engaging in many contexts. As [Ran] Liu [chief AI scientist at Amira Learning] explained, children who believe themselves to be behind are the least likely to engage in these settings. Interfaces that leverage AI can offer constructive feedback that does not carry the same stakes or cause the same self-consciousness as a human's response. Learners are therefore more willing to engage, take risks, and be vulnerable.[51]

This creates an entirely new dynamic to the education system. Knowing that your teacher, educator, or evaluator is not a human greatly reduces or eliminates any level of judgment and minimizes (hopefully) biases that may inhibit one's full capacity to learn. Clever, some may think. No, cold and distant for others. But for some learners, this is exactly what they need.

And finally, one last aspect of education to which AI can assist involves the capacity with which it can single out and identify specific needs of individual learners quicker and perhaps more effectively than a single teacher. There is widespread agreement in the education field that the inability to judge a learner's skill profile is a leading industry challenge. What the attendees found was that AI has the potential to quickly determine a learner's skills, recommend solutions to fill the gaps, and match them with roles that actually require those skills.[52] Imagine a teacher with a classroom of thirty students who must expend a considerable amount of energy trying to best understand the learning needs, abilities, and shortcomings of each of her students. Now, imagine an AI application that can do this quicker, more efficiently, and more effectively, which also frees up a teacher's time to focus on other, more pressing areas of concern. Just one example of this would involve students who are learning the English language. Since there will be differences in abilities to speak, write, and converse in English, knowing precisely what a student's abilities and weaknesses are in this one area will allow teachers to assign work more accurately and assess their students accordingly. AI-powered adaptive tutoring systems can provide real-time assistance to students, offering additional support and guidance in areas where they may be struggling. These systems can adapt to students' progress and adjust the difficulty of tasks accordingly. AI-driven adaptive learning platforms use data analytics to identify students' strengths and weaknesses, tailoring content and learning experiences to optimize educational outcomes.

The U.S. Department of Education has also released its recommendations for the use of AI in education. Touching on many of the same findings of the Stanford Summit, the Department of Education recommends keeping humans deeply entrenched in the loop—that is, AI should never replace teachers. It should only enhance their teaching by facilitating their abilities to educate. Additionally, the department recommends that all AI models must align with a shared vision for education.

> We call upon educational decision makers, researchers, and evaluators to determine the quality of an educational technology based not only on outcomes, but also based on the degree to which the models at the heart of the AI tools and systems align to a shared vision for teaching and learning.[53]

AI tools can automate routine administrative tasks such as grading, scheduling, and record keeping, allowing educators to focus more on teaching and mentoring students. These systems can adapt to students' progress and adjust the difficulty of tasks accordingly. And, just like the printing press, AI continues to contribute to breaking down geographical barriers by providing access to educational resources and courses to individuals worldwide, especially through online platforms. And lastly, AI can support educators in their professional development by offering personalized training programs, suggesting relevant resources, and keeping them updated on the latest educational trends.

Further Applications of AI

There are many other areas within society where AI will bring greater efficiency, sustainability, and hope for a future with greater potential and possibilities. Here are some examples:

- AI will be able to enhance security systems by detecting anomalies in data patterns, which will help in preventing cyberattacks. It can also be applied in military settings for threat analysis and strategic decision-making.
- In terms of environmental sustainability, AI can be applied to optimize resource consumption, energy efficiency, and waste management. Smart grids, for example, can balance energy supply and demand, reducing environmental impact while greater solutions to existing problems like plastic contamination in the oceans or particulate contamination in the air can be brought about and deployed in time-, money-, and energy-saving manners. Determining better, more energy-efficient ways to live will be a boon not only to human inhabitants but to all species affected within connected ecosystems.
- We have witnessed how much AI has assisted in the medical sciences; it will also assist greatly in other areas of science as well. For instance, in analyzing data, AI programs can process and analyze vast datasets, accelerating scientific research in fields such as astronomy, physics, and climate science. It can help researchers identify patterns and trends that may be challenging for humans to discern. And it can do it thousands of times quicker, with greater accuracy, and more efficiently. This will speed up all processes in scientific investigation and discovery.

- And in terms of transportation, AI will help optimize traffic flow to decrease traffic jams and get people to their destinations faster. In 2016, I met with high-level staff of the Ontario Ministry of Transportation to discuss my proposal for relieving congested traffic on the 400-series highways in and around the Greater Toronto Area (or GTA). I presented to them the idea of using John Nash's understanding of algorithmic behavior of birds in flight and how this could be transferred to rush hour traffic. If drivers abided by three basic principles or rules* at the start of bottleneck rush hour traffic, all traffic would generally flow more efficiently—but the trick is this: everyone must cooperatively abide by the three basic principles. It is these same algorithmic principles that autonomous vehicles will use to maximize efficiency and decrease congestion. But notice how autonomous vehicles literally take humans out of the decision-making equation? With my proposal, humans had to cooperate collectively for it to work. With autonomous vehicles, human agency is taken entirely out of the control of the vehicles, which, in turn, makes them all work much more efficiently. Individual biases are supplanted with cold, calculated, precision—the very DNA of machine optimal functionality. And so the collective group will largely win at the minimal sacrifice of individual liberty. Long gone will be the days of idiot highway drivers zipping in and out of traffic; jeopardizing their lives and the lives of those around them. Gone will be the tailgaters, the preoccupied, the texters, and the drivers in the fast lane who are going way too slow; they will all be gone in the lane for autonomous vehicles. And it will hopefully be the fastest, safest, and most efficient lane on the highway.

Part II: Requests to an AI God: Speculations on the Benefits of AGI/ASI

If you were given the chance to be granted wishes from a future, very powerful AI, what would they be, and why? Like someone with the lamp of a genie, wish well—lest you choose unwisely and release a Pandora's Box of unintended consequences upon the world. There are many ways in which our lives will be enhanced through the development of advanced AI technologies. But the main desire we all commonly want satisfied with such

---

* 1. Always match the speed of the vehicle in front of you. 2. Always stay two car lengths behind the vehicle in front of you. 3. Don't change lanes (until you need to exit from the highway).

a powerful AGI/ASI god is the very same desire we have for basic ANI technology: the desire to increase our comfort and pleasure levels and/or decrease our levels of discomfort and displeasure. We have evolved just as all other mammals have: with the propensity for survival and reproduction. As such, we possess nervous systems and brains that are hardwired and conditioned to respond to pleasing stimuli while avoiding consequences that cause undue pain or discomfort. It follows, then, that the main purpose of such an AI god will be to service our desires to increase our pleasure and decrease our pain. And this could come in any number of forms.

We Are All on Drugs

Humans like to get high. Whether it's through artificial or natural means, humans generally do things that make them feel better—either by elevating their mood, or diminishing their pain and discomfort. When you were a child, you may have enjoyed amusement park rides, vacations with family, playing with friends, and enjoying thousands of other activities from the mundane to the sublime that made you feel good or happy or elevated your mood in some positive way. With each experience, you were getting high. That means the chemical neurotransmitters in your brain were coursing through your neural synapses and making you feel happy, excited, and exhilarated. Humans, like all mammals, tend to gravitate toward situations that elevate their mood and avoid those situations that cause discomfort. It's basic biology. Our brains are hardwired to increase our likelihood to survive and reproduce. That's why eating and sex are so enjoyable. They both satisfy the two most fundamental drives of the mammalian world: surviving and reproducing.

But there are substances that contain chemicals that shortcut the evolutionary process, and provide us with amusement, enjoyment, and happiness without feeding us or satisfying us sexually. Drugs and alcohol provide enough artificial stimuli in our lives to allow us to use them to the point of abuse. And let's face it; there's a lot of drug abuse out there. Alcohol is clearly the largest, most abused drug throughout the world. But there are also opioid epidemics going on throughout different parts of the world as well. Drugs and alcohol in their various forms have been used for centuries by humans for celebration, inspiration, and general intoxication. But they've also ruined and continue to ruin far more lives than any other substances throughout history. Right now, the world could use at least one new and better recreational drug; in other words, the world needs to get high, better. And such a god could easily make this happen.

So, if we're going to continue to get high, maybe we should learn a lot more about the processes, mechanisms, pathologies, and many other factors behind our desire for such mental states of being. And perhaps, AI will be able to assist in making very accurate models of brain behavior and drug interactions within our brains so that scientists will be able to produce substances whose rewards do not also cause such devastating effects. So one of our challenging requests posed to the new god of AI might be something like this:

> Create a new recreational drug or neural stimulant that makes us feel good without any side effects, does no internal tissue damage, is non-addictive, has no lasting harmful effects, leaves no next day hangovers, and whose effects can be neutralized if consumed accidentally by those for whom it is not intended (e.g., children, pets, etc.).

Too many lives are being destroyed by alcohol, opioids, methamphetamines, and other such nasty substances. Since humans like to get high—and they're not going to stop wanting to get high—we need to consider seriously how AI might be able to help in this regard. I think it is relatively safe to say that, with greater knowledge of what stimulates human brain activity related to reward, pleasure, alleviation of pain and discomfort, and such, the more humans will gravitate toward it. Therefore, it follows that we will (or, at least, should) utilize the various benefits of AI advancement to better understand the safest, yet most effective, way to get humans high.

Your Systemic Self

But what if our AI god could simply bypass the crude usage of such chemical interactions and simply found better ways to give us pleasure and avoid pain? Just consider our "selves" for a moment, purely as biological beings, and how our understanding of and treatment of health matters might change with such a god. For example, imagine if an AI god agreed to help and not harm us and knew all possible relevant health information about you. That is, it knew your entire genotype, biome, neural brain structure, circulatory system, endocrine system, digestive system, every organ, and all your other parts, systems, or processes that make you, well . . . you. It would have very precise information regarding the construction, function, and processes of what I refer to as "your systemic self."[54] It would then be able to diagnose and assess all of your systems and be able to spot potential diseases before they became problematic. Recently, a team of scientists in

Denmark have developed something called the "doom calendar," which can roughly predict the date of a person's death.

> [T]he model, called life2vec, crunched data—age, health, education, jobs, income and other life events—on more than 6 million people from Denmark supplied by the country's government, which collaborated on the research. . . . As life2vec evolved it became capable of building "individual human life trajectories.". . . "The whole story of a human life, in a way, can also be thought of as a giant long sentence of the many things that can happen to a person," the paper's author, Sune Lehmann, a professor of networks and complexity science at the Technical University of Denmark, said. . . . Eventually, the AI construct was able to correctly predict those who had died by 2020 about 78 percent of the time, researchers say in the report.[55]

Although none of the participants were told what their death predictions would be, would you be interested in finding out what your future holds in store for you? This type of relational systemics is about to become a lot more common in medical practice and elsewhere.

But an AGI/ASI god could do much more than simple, probabilistic predictions about your general finitude. It could also potentially cure your ailments or supply you with the necessary mechanical prosthetics or bionics you needed to replace your worn out parts. Or it might simply make a cloned copy of you and create a younger, better, healthier version of yourself while transferring your consciousness into the newer, more improved model of you—delaying sickness, decrepitude, and even death for decades, centuries, or even longer. Or it could simply do away with fragile things like human bodies and make an exact digital replication of your brain, upload you to a mainframe—where you could live out your wildest dreams in an infinite virtual reality. Or it could download your digital brain into an autonomous android where you can stay in the so-called physical world and live out your life eternally and freely with virtually no pain and with powers and abilities far greater than you could ever possibly attain as a "natural" human.

Are you ready for such a future?

Or do these prospects frighten the bejesus out of you? I do not intend to falsely startle or upset you, dear reader. But we do need to get ready—because we are not that far away from a very specialized form of medical care in which the commonalities of the human condition, combined with the peculiarities of each one of us, will lead to a very focused and specialized form of medical care like we've never seen before. Instead of being diagnosed with a specific condition and given a blanket amount of medication, the future of medicine will be so precise, it will determine much more accurately and effectively not simply what we need to do, but by how much. And there will be options available that currently seem like magic and science fiction.

Keep in mind that, in the next chapter, we will consider ways in which AI advancements may exploit this human desire for pleasure and absence of pain through various forms of manipulation. In the movie *The Matrix*, the betraying character, Cipher, made a deal to have his consciousness transferred back into the virtual program, have his memory scrubbed, and be granted a good position in life with plenty of resources—after, of course, he turns his friends over to the cyberfeds. To Cipher, reality became too much to bear. He became tired of the rebellious fight and was looking forward to living in an indistinguishably simulated unreal world again. As AI continues to become more and more powerful, will it reach a point at which it could easily manipulate us into doing what it wants? If it could move money or resources around quite effectively, what deals might we find irresistibly alluring when given the opportunity from an AI-generated deal maker? Who might we be willing to sell out to such alluring and enticing offers in the name of our own self-preservation? Or, perhaps, the preservation of our family? Or, say, for infinite pleasure and immortality? As good as AI will become and will bring us improvements in our daily lives, so too will it present ethical issues and dilemmas. For not all AI developments will be positive. It is toward the potential negative aspects of AI that we now turn our attention.

The Risks and Harms of AI   3

With all new forms of technologies there are positive and negative effects; we have experienced this with the advent of automobiles, air travel, nuclear power, and the internet. And many such discoveries have produced dual-use technologies (i.e., technologies used by both civilians and the military). The same holds true for artificial intelligence. Now that we have seen just a few of the ways AI advancements are going to greatly benefit humankind, we can reflect on the potential negative effects by considering the risks and harms that are heading our way as we advance forward with ever-expanding and transformative AI technologies.

Many of us are already quite familiar with the social consequences of AI algorithms manipulating human behavior, for example, spreading of misinformation and disinformation, shaping public opinions, spreading conspiracy theories, and impacting democratic processes.[1] This will continue to worsen as more and more sophisticated technologies allow users to create, distort, and alter information. The absence of truthfulness in generative AI systems, producing content (images, texts, audios, videos) that lacks correspondence with reality but appears convincingly genuine, will be used intentionally by those who prey upon the susceptible—such as the elderly—and by hackers interested only in conning people out of their money.

Additionally, we have seen, to some degree, the perpetuation of stereotypes, discrimination, and oppression through algorithmic biases. We have experienced this in specific AI programs where, for example, facial recognition technologies across a wide array of algorithms have been least accurate on women of color.[2] Similar biases could arise in other judgments—for example, in the analysis of candidates for financial loans, job interviews, and such.

Another concern with AI technologies is the lack of transparency in both the use and design of AI models. Often referred to as the "black box problem," there are two issues at play here: many creators of AI technologies keep quiet regarding the specifics behind the technologies of their products in order to protect their trade secrets and patents. But there is something much more pernicious at play within that "black box": the creators of such systems have little idea how such deep learning systems come to their conclusions. According to Elliot Mckernon:

> Of course, AI engineers understand how to set up, train, and deploy these AI models, but during training the models become so complex that it becomes impossible to understand what's going on under the hood. This is analogous to the human brain: we can understand the overall structure of the brain and we can understand how individual neurons work, but we can't look at someone's brain and explain why they made a particular decision.*

Not knowing how such systems "think" or make "decisions" translates into trouble; for they can produce unwanted outcomes—like autonomous devices acting in unpredictable ways. When they behave irregularly, there is no way to investigate as to why; and if we cannot determine why a system acted in such a way, then we lack the ability to predict with certainty if and when it will act that way again.

And then there is the problem of copyright infringement and the unauthorized use of extensive training data without the consent of or compensation for its creators. When ChatGPT writes you up a nice synopsis regarding some topic, where did that information come from? What actual living, breathing, persons were responsible for it? And how do they get compensated—if at all? Battles are currently being waged by people like Margaret Atwood and Sarah Silverman over the copyright of such content and how this can be usurped by such new technologies.

In reference to environmental sustainability, many are calling out the hefty environmental impact and carbon footprint of data centers and neural networks required to run such AI systems. The computing power needed for these systems is immense and requires a great deal of energy; this can translate to a considerable drain on current electrical systems.

---

* Mckernon provided this excellent feedback while reading through the manuscript.

In January [2024], the International Energy Agency (IEA) issued its fore-cast for global energy use over the next two years. Included for the first time were projections for electricity consumption associated with data centers, cryptocurrency, and artificial intelligence. The IEA estimates that, added together, this usage represented almost 2 percent of global energy demand in 2022—and that demand for these uses could double by 2026, which would make it roughly equal to the amount of electricity used by the entire country of Japan.[3]

While it is true that such data centers consume huge amounts of energy, some are proposing ways to deal with this:

[I]t looks like Microsoft is betting on advanced nuclear reactors to be the answer . . . for all aspects of nuclear energy infrastructure for global growth. Microsoft is specifically looking for someone who can roll out a plan for small modular reactors (SMR). All the hype around nuclear these days is around these next-generation reactors. Unlike their older, much larger predecessors, these modular reactors are supposed to be easier and cheaper to build.[4]

Although Bill Gates is a huge fan of small modular reactor designs and founded and chairs TerraPower—a company devoted to developing such reactors—there are currently no plans to sell any to Microsoft. Instead, Microsoft already has a deal to buy clean energy credits from Canadian utility Ontario Power Generation, which is on track to be the first utility to deploy an SMR in North America.[5] And we may want to look to the future with the hope that AI advancements in scientific technology might help with developments in other forms of cleaner, cheaper, energy—like nuclear fusion.

Aside from environmental concerns, there is also considerable appre-hension regarding the potential for AI to displace millions of jobs. Whether it's service industries cutting workers for greater automation or white-collar workers in finance and banking who will be replaced with ever-improving predictive software, it is clear that the economies of the world are about to change—some for the better; but some for the worse. This leads to the consideration of a potentially bigger problem: the concentra-tion of power in the hands of a few controlling the current AI systems, leading to an oligopoly and limited diversity in decision-making. We are already seeing the biggest companies like Google, Microsoft, OpenAI,[6] and others competing at breakneck speed to be the first to capitalize on the applications of new forms of AI technology. Who will take the lead? And to what extent can some businesses compete without the great advancements of such new technologies?

Another potential harm involving the development of novel AI technologies involves the effects it will have on mental health. People with anxiety are more likely to become fearful of the potential risks and harms that may result from AI systems. The heightened fears of potential risks will increase global anxiety and uncertainty and produce a similar, but potentially worse fear and paranoia than that witnessed regarding nuclear proliferation during the Cold War years. Simultaneously, some will become overdependent on conversational or interactive robotic artificial intelligence agents, while others will be kept further isolated from friends and family. If lifelike robots seem to genuinely care about you and look after your needs—*all* of your needs—will the bonds among family, friends, and lovers become weakened and atrophy?

And finally, the greatest concern with emerging AI technologies involves the potential for existential risk (or x-risk). There is considerable potential for catastrophic harm when building a system with intelligence capabilities far surpassing our own. We have no idea how such a system will respond to our attempts for moral alignment, control, or containment. At this point in time, there is a wide spectrum of people ranging from those who maintain that very little harm will result from such AI technologies to those who believe that AI will cause the end of human civilization as we know it. The spectrum spans from naysayers to doomsayers or from Y2K to Armageddon. We must face the reality of considering how we should develop such a powerful form of technological advancement. Even if the existential risk is only small, say 5 percent,* we have a moral obligation to ourselves, future generations, other species, and the planet to ensure that such a powerful form of AI *never* escapes our control. Let's take a look at these potential negative effects of AI in greater detail.

So . . . What Could Go Wrong with AI, and How Might This Happen?

Rational versus Moral Agency

It is important to realize that there are several categories that define how AI might present risks and generate harm. The first category involves the concept of "agency." This concept is further distinguished by two

---

* And some would say a chance of x-risk even as low as 0.01 percent is something we should take quite seriously.

subtypes: rational agency and moral agency. An agent is considered "intelligent" or "rational" if it acts in a way that "maximizes the expected value of a performance measure based on past experience and knowledge."[7] Lin Padgham and Michael Winikoff both maintain[8] that an agent is intelligent if it is situated in an environment and responds in a timely manner to changes in the environment, proactively pursues goals in a flexible and robust way, and is capable of belief-desire-intention analysis. But others, such as Andreas Kaplan and Michael Haenlein, provide accounts of agency that mirror definitions of artificial intelligence itself, in other words, as "a system's ability to correctly interpret external data, to learn from such data, and to use those learnings to achieve specific goals and tasks through flexible adaptation."[9] In this context, a Roomba could be considered a rational agent; however, a Roomba has extremely limited so-called intelligence. What really concerns us is whether or not, as a rational agent, a much more advanced form of AI technology could cause harm simply in pursuit of its goals.

What also concerns us moving forward is whether or not, and to what degree, AI technology could develop "moral agency." A misaligned, rational AI agent can generate risk and cause harm without having any concept or understanding of the harm it is inflicting. Hence, it is amoral. And only moral agents are held accountable for their actions. But to be held morally accountable requires more than simply being a rational agent. In this context, "moral agency" refers to a number of factors such as sentience, self-awareness, and the acknowledgment that one's actions have consequences relative to a proposed system of value. At this point in history, there has been no clear indication that any AI developments possess moral agency.* This means either that

- humans who do possess moral agency will deliberately direct the harmful results of AI technology;
- the AI itself could develop moral agency and decide to act of its own will and volition to generate harm through some system of value; or
- the AI will possess only rational agency and simply become misaligned and will functionally pursue goals that will bring about harmful consequences.

---

* Of course, Blake Lemoine, a Google AI engineer, has claimed LaMDA is sentient and possesses moral agency. But this is still quite debatable.

But always keep this in mind: an AI technology need not possess moral agency to generate harm; plenty of AI technologies lacking moral wills or volitions could generate harm without realizing it. And we will consider a few scenarios in which this could happen in the final section of this chapter.

Intentional versus Unintentional Harm

The second category in which AI developments might generate risk or harm involves the concept of "intent"—that is, either risk or harm from AI could be caused *intentionally* or *unintentionally*. For example, some person, company, or country might *intentionally* make a deliberate decision to use the advances of AI to generate risk and harm to others; or they might do so *unintentionally*—perhaps accidentally. The same can be said for AI itself—that is, as more advanced forms of AI develop, the more autonomous they will become. The University of British Columbia's Centre for Artificial Intelligence Decision-Making and Action maintains that AI systems for decision-making can be understood as lying along a spectrum according to their levels of autonomy. In some cases, human experts use AI techniques to support them in reasoning about a single, high-stakes decision.[10]

In some cases, humans are primarily involved in decision-making situations (e.g., environmental, military, etc.). In other cases, there is a blended approach whereby AI is more actively involved in a mixed initiative (e.g., medical diagnostics, educational instruction, etc.). And in other settings, it makes sense for an AI system to make autonomous decisions (e.g., auto-pilots). In each of these cases, such forms of AI, therefore, might generate harm either intentionally or not—whether they possess moral agency or not.

Misuse versus Misalignment

And the final category considers to what extent such risks and harms were the result of either *misuse* or *misalignment*. In other words, either AI technologies will be *misused* at the hands of humans, or the commands we program into the AI will become *misaligned* with what we want it to do and generate harm in some form.

To clarify, consider the three categories like this:

1. Either current and future forms of AI technologies will generate harm at the hands of some *moral agent* such as an individual,

a group of people, a company, corporation, city, or country; or some AI technology, of its own doing, will generate harm accidentally and without *moral agency*.

2. Some moral agents may choose to use AI technologies to deliberately harm others *intentionally*. But some nonmoral agents—like unconscious AI—might generate harm *without intending* to do so. If a form of AI should develop moral agency (i.e., through conscious awareness), it may choose to cause harm; or it may not. We simply don't know.

3. And finally, when a moral agent—like a human—intentionally sets out to harm some person or group through AI technologies, they are *misusing* it. But in harming others, a form of AI technology may simply be *misaligned* with the moral rules to which we program it to comply. In other words, we have no idea to what degree AI will comply or fail to comply (i.e., align with our moral precepts). It may not intentionally choose to act immorally or unethically; its harmful actions may simply be the result of its inability to align with our ethical precepts.

This last point refers to what is known as the alignment problem, which, according to Stuart Russell and Peter Norvig, can be defined in the following way:

> In the field of artificial intelligence (AI), AI alignment research aims to steer AI systems toward a person's or group's intended goals, preferences, and ethical principles. An AI system is considered aligned if it advances its intended objectives. A misaligned AI system may pursue some objectives, but not the intended ones.[11]

The difficulty in keeping a superintelligent AI aligned emerges because there is such a wide range of both desired and undesired behaviors to monitor. Aligning AI involves two central challenges: the first involves what is called "outer alignment," or carefully specifying the purpose of the system, while the second is "inner alignment," or knowing that the system has fully adopted its purpose faithfully. To help with this, designers will sometimes use "'proxy goals," which require human approval before being undertaken by the AI system. But even precautions such as this can be bypassed if unsuspected loopholes arise, or the AI itself is deceiving its users into believing that it is aligned.

Misaligned AI systems can malfunction and cause harm. AI systems may find loopholes that allow them to accomplish their proxy goals efficiently but in unintended, sometimes harmful, ways (reward hacking). They may also develop unwanted instrumental strategies, such as seeking power or survival because such strategies help them achieve their final given goals. Furthermore, they may develop undesirable emergent goals that may be hard to detect before the system is deployed and encounters new situations and data distributions.[12]

Researchers suggest that aligning advanced AI systems in the future could pose major challenges because, as they become more powerful, they might figure out ways to manipulate their rules or "game" their specifications by finding loopholes, tricking their creators, and growing even stronger by seeking power in ways that are not penalized. This could lead to more serious consequences. Additionally, since these AI systems will likely be more intricate and independent, understanding and controlling them could become even more difficult.

At this particular point in history, although we are well on our way to building a superintelligent machine god (SMG), only artificial narrow intelligence (ANI) exists, and as such, it cannot possess moral agency. This means that, currently, the likelihood for AI risks and harms lies in the potential for two outcomes:

1. Human *moral agents* may *intentionally misuse* ANI technologies to the harm or detriment of others.
2. Artificial *non-moral agents* may *unintentionally* cause harm due to moral *misalignment.**

As the advancements toward building an SMG increase, we will need to revisit its capacity to think and act as a moral agent. For now, as we saw earlier in the chapter, there are a number of ways in which ANI technologies pose risks and potential harms. Let's examine some of these more carefully.

---

* Of course, there still lies the potential for a nonagent AI to develop agency through some form of conscious awareness. Should any form of AI become conscious in manners similar to those of any human, it would automatically, immediately be attributed the same moral and legal rights as any human. And as such, it would possess the capacity to misuse its attributes in the generation of harm.

Risks from Current ANI Systems

Manipulation through Misinformation/Disinformation

The rise and proliferation of AI-generated content, including deepfakes,* plays a role in propagating falsehoods and manipulating public sentiment. With the past pandemic, we saw an unusually high level of conspiracy theories circulating online. In fact, in my own research, I've witnessed a significant spike in conspiracy theory popularity—especially with the rise of groups like QAnon.[13] With new AI technologies, the ability to send out false, misleading disinformation will become not only much easier but also will be much more convincing. The elderly, the digitally unsavvy, children, and millions of others will be exploited by the spread of such falsehoods.[14] It is crucial to undertake significant endeavors to detect and combat AI-generated misinformation, as it is vital for safeguarding the integrity of information in the digital era. This has led to an inevitable arms race between the use of AI for the creation/distribution of false information and the media resources available to combat it and educate the public about it.

In an excellent article, Alexandra Samuel makes a very good case for how we could learn from our mistakes with social media in order to do a better job with "getting AI right."[15] Her position focuses on how greed led to the drive for advertising dollars with social media platforms. As L. M. Sacasas states: "[D]igital platforms driven by ad revenue models were designed for addiction in order to perpetuate the stream of data collected from users."[16] This perpetuated the use of content that was intended to be alarming and sensational both to those on the political left and right. This, in turn, continued to divide both sides on heated issues, which deepened the capacity for algorithms to match content to individual biases leading to deeper confirmation-biased entrenchment of a political stance. Samuel maintains that much of the terrible, destructive impacts of social media follow this type of core dynamic:

> The bite-sized velocity of social media has made it endlessly distracting and disruptive to our families, communities, relationships, and mental health. As an ad-driven, data-rich, and sensational medium, it's ideally suited to the dissemination of misinformation and the explosion of anti-democratic manipulation. And as a space where users create most content for free, while companies control the platforms and the algorithms that determine

---

* A video of individuals that did not film the actual persons presented, that is, they were artificially generated.

what gets seen, it has put creators at the mercy of corporate interests and made art subservient to profits.[17]

For those of you who have seen the documentary *The Social Dilemma*, you will recall how Tristan Harris stated quite frankly how social media algorithms configure your data with considerable accuracy to create your digital persona. This persona can then be manipulated according to how information becomes directed at it.

> When you look around you, it feels like the world is going crazy. You have to ask yourself, like, "Is this normal? Or have we all fallen under some kind of spell?"
>
> I wish more people could understand how this works because it shouldn't be something that only the tech industry knows. It should be something that everybody knows.[18]

Others in the tech industry knew something else was going on with various platforms. Social media critic Jaron Lanier says:

> It's the gradual, slight, imperceptible change in your own behavior and perception that is the product. And that is the product. It's the only possible product. There's nothing else on the table that could possibly be called the product. That's the only thing there is for them to make money from. Changing what you do, how you think, who you are. It's a gradual change. It's slight. If you can go to somebody and you say, "Give me $10 million, and I will change the world one percent in the direction you want it to change." It's the world! That can be incredible, and that's worth a lot of money.[19]

Everything you do online, on your cell phone, on your home computer, and through Siri, Alexa, or whatever, ends up being monitored in some way by some tracking algorithms that then provide your information, your data, to companies who will use it to learn more about human beings as the algorithm functions online. The more data they collect, the more they can accurately direct marketing information toward you, or send you stories that confirm your biases in politics, fashion, sports, or just about any facet of the human experience. This is often referred to as "surveillance capitalism" (i.e., the profiting from the infinite tracking of online behavior by Big Tech companies). Tristan Harris sums it up by saying this:

> At a lot of technology companies, there's three main goals. There's the engagement goal: to drive up your usage, to keep you scrolling. There's the growth goal: to keep you coming back and inviting as many friends

and getting them to invite more friends. And then there's the advertising goal: to make sure that, as all that's happening, we're making as much money as possible from advertising. Each of these goals are powered by algorithms whose job is to figure out what to show you to keep those numbers going up.[20]

One of the most concerning aspects of the way in which online social media platforms work is the manner in which they all collectively cater to you as an individual "digital self." And if you thought that others view the exact same information as you; or you, theirs, you'd be wrong:

> We have all been living under the delusion that our digital lives online are roughly the same as everyone else's. But it is quite the opposite. And every time we search our preferred websites, we get more information which satisfies our confirmation biases, which, in turn, provide us with neural payloads in the form of raised levels of neurotransmitters like dopamine, serotonin, and oxytocin. So, every time we confirm our already established biases regarding any particular subject, we feel good because we're getting a little high. And those major social media companies producing the content know and exploit this algorithmic–neural relationship very well.[21]

I have worked for decades in the interdisciplinary fields that study the relationships between ideas and neurological reward.[22] Because marketers with Big Tech platforms can gather enough information about your online habits that elevate your neurological payloads, they will exploit it. The field is called neuroeconomics.

Now think for a moment; if this has already been happening without the use of advanced AI technology, we can only imagine what the future spread of misinformation, disinformation, and conspiracy theories will look like as it becomes ever more prevalent in our lives. Samuel warns, if we're not careful, we're going to fall for the same trappings of Big Tech social media platforms:

> We're embracing technologies that create content so rapidly and so cheaply that even if that content is not yet quite as good as what humans might create, it will be more and more difficult for human creators to compete with machines. We're accepting opaque algorithms that deliver answers and "information"—in quotes, because AIs often present wholly invented "hallucinations" as facts—without much transparency about where this information came from or how the AI decided to construct its answers.[23]

Samuel advocates for a type of steadied maturity in how we proceed with AI developments. She believes that "those of us who are truly inspired and enchanted by the advent of new technologies are the ones who most need to rein in our enthusiasm; to anticipate the risks and to learn from our past mistakes."[24] And this also means that it's Big Tech who must take the initiative to invite regulation and welcome transparency in their efforts moving forward. But she warns that regulators should not depend on the technical advice of AI executives in order to set appropriate rules. This is because "even well-intentioned execs are going to be less than objective about regulations that constrain their potential for profit."[25] Instead, she applauds Meg King and Aaron Shull who maintain that "policy makers must prioritize developing a multidisciplinary network of trusted experts on whom to call regularly to identify and discuss new developments in AI technologies, many of which may not be intuitive or even yet imagined."[26] Samuel echoes many of the same sentiments that my colleagues and I at Convergence Analysis, as well as other AI governance organizations, hold dearly: "It's going to take international coordination and investment to develop an independent source of regulatory advice that is genuinely independent and capable of offering meaningful advice: Think of an AI equivalent of the World Health Organization, with the expertise and resources to guide AI policy and response at a global level."[27]

In the final chapter of this book, I offer ways in which readers can empower themselves with knowledge and resources regarding AI development and advancement. Samuel believes that we should not be passive but proactive in learning more and more about the potential of AI technologies:

> [L]et's figure out how to be the agents who use the tools, rather than the subjects who get manipulated. We won't get there by avoiding Chat-GPT, DALL-E and the like. Avoidance only makes us more vulnerable to manipulation by artificially generated content or to replacement by AI "workers." Instead, we human workers and tech users need to become quickly and deeply literate in the tools and technologies that are about to transform our work, our daily lives, and our societies—so that we can meaningfully shape that path. In a delightful paradox, the AIs themselves can help us achieve that rapid path to AI literacy by acting as our self-documenting guides to what's newly possible.[28]

Using AI to better understand and adapt to AI is sound advice. Making the unfamiliar and daunting accessible and understandable is a proactive way in which the public can keep up with the developing AI technologies as they appear in real time.

Some scholars are worried about what may happen to elections when AI technologies become more and more sophisticated and capable of manipulating human behavior. Professors Archon Fung and Lawrence Lessig have proposed a scenario in which an AI system will interfere in any election process to the point of changing people's voting behavior. Dubbed the "Clogger," Fung and Lessig's machine would relentlessly pursue a single objective: "to maximize the chances that its candidate—the campaign that buys the services of Clogger, Inc.—prevails in an election."[29]

> As a political scientist and a legal scholar who study the intersection of technology and democracy, we believe that something like Clogger could use automation to dramatically increase the scale and potentially the effectiveness of behavior manipulation and microtargeting techniques that political campaigns have used since the early 2000s. Just as advertisers use your browsing and social media history to individually target commercial and political ads now, Clogger would pay attention to you—and hundreds of millions of other voters—individually.[30]

Fung and Lessig believe that the process to change your voting behavior involves three steps:

1. Its language model would generate messages—texts, social media, and e-mail, perhaps including images and videos—tailored to you personally. Whereas advertisers strategically place a relatively small number of ads, language models such as ChatGPT can generate countless unique messages for you personally—and millions for others—over the course of a campaign.
2. Clogger would use a technique called reinforcement learning to generate a succession of messages that become increasingly more likely to change your vote. Reinforcement learning is a machine-learning, trial-and-error approach in which the computer takes actions and gets feedback about which works better in order to learn how to accomplish an objective. Machines that can play Go, Chess, and many video games better than any human have used reinforcement learning.
3. Over the course of a campaign, Clogger's messages could evolve in order to take into account your responses to the machine's prior dispatches and what it has learned about changing others' minds. Clogger would be able to carry on dynamic "conversations" with

you—and millions of other people—over time. Clogger's messages would be similar to ads that follow you across different websites and social media.[31]

What is interesting about this proposed scenario is that it's not simply a philosophical thought experiment. Instead, it's a valuable tool that is used today in anticipating how emerging AI technologies may affect us socially. The sociotechnical concerns of the effects of AI are still very much a part of a world that Norbert Weiner envisioned in the 1950s.

Fung and Lessig even go so far as to imagine that, should competing parties use such devices in the future,

> [t]he president will have been elected not because his or her policy pro-posals or political ideas persuaded more Americans, but because he or she had the more effective AI. The content that won the day would have come from an AI focused solely on victory, with no political ideas of its own, rather than from candidates or parties. In this very important sense, a machine would have won the election rather than a person. The election would no longer be democratic, even though all of the ordinary activities of democracy—the speeches, the ads, the messages, the voting and the counting of votes—will have occurred.[32]

And if winning politicians continued on this trajectory, they would simply become more and more reliant on the Clogger to manipulate the public further so they could stay in power longer. When we consider the topic of governance in chapter 5, we will look carefully at what is happening in the world, right now, regarding the ways in which such futures can be avoided, and democracy, as we have known it, can be preserved. At this point in time, some nations have already taken steps in acknowledging the seriousness with which people could be manipulated by AI-generated technologies in the future. Fung and Lessig note that European Union (EU) regulators are moving in this direction. Policymakers revised the European Parliament's draft of its Artificial Intelligence Act to designate "AI systems to influence voters in campaigns" as "high risk" and subject to regulatory scrutiny.[33] It is important to note that the threats of spreading false or misleading information can be combatted best through media and online literacy. The more familiar we become of how misinformation and disinformation can be spread, the better off we'll be. To be forewarned is to be forearmed, as they say.

BUT WHAT HAPPENS IF THE AI GOES BOTSHIT CRAZY AND MISINFORMS US ON ITS OWN?

I might as well keep you up to speed now, dear reader, and let you know well in advance that deception in AI is one of the biggest concerns we have to consider. Many of you will already be quite familiar with how various large language models (LLMs) like ChatGPT can "hallucinate" or produce what some now call "botshit."[34] This is the creation of false or imagined information from the AI source itself.

> In a recent paper, Tim Hannigan, Ian McCarthy and [André Spicer] sought to understand what exactly botshit is and how it works. It is well known that generative AI technologies such as ChatGPT can produce what are called "hallucinations." This is because generative AI answers questions by making statistically informed guesses. Often these guesses are correct, but sometimes they are wildly off. The result can be artificially generated "hallucinations" that bear little relationship to reality, such as explanations or images that seem superficially plausible, but aren't actually the correct answer to whatever the question was.[35]

André Spicer and his colleagues maintain that such forms of generative AI could threaten the very fabric of democracy itself:

> [T]here is a real danger it could create an environment where some people start to make important voting decisions based on an entirely illusory universe of information. There is a danger that voters could end up living in generated online realities that are based on a toxic mixture of AI hallucinations and political expediency.[36]

But there is a general consensus among those who work in AI that we are not powerless against such potentialities. Spicer and his colleagues suggest a number of ways in which we can be forewarned and forearmed against the coming wave of AI developments. A white paper has been produced by a group of scholars from the University of Chicago and Stanford University, which outlines how the public could have prepared itself better for generative AI in the 2024 election.[37]

> Although AI technologies pose dangers, there are measures that could be taken to limit them. Technology companies could continue to use watermarking, which allows users to easily identify AI-generated content. They could also ensure AIs are trained on authoritative information sources. Journalists could take extra precautions to avoid covering AI-generated stories during an election cycle. Political parties could develop policies to prevent the use of deceptive AI-generated information. Most importantly,

voters could exercise their critical judgment by reality-checking important pieces of information they are unsure about.[38]

At this point in history, we should remain cautiously optimistic that our values of fair and competently regulated and monitored elections will thwart the attempts of those who would use AI to interfere with such a valuable civic process. The Brookings Institution has issued a commentary[39] on how voting officials, election administrators, and the public can empower themselves with information from reliable sources:

> AI-friendly and -wary lawmakers alike have begun setting their sights on a bipartisan AI regulatory regime. And civil society has an important role to play, demonstrated by the work of organizations like the Leadership Conference on Civil and Human Rights, the Lawyers Committee for Civil Rights under Law, the Brennan Center for Justice, Public Citizen, and Brookings (which does its own extensive work on the subject). They, along with many others, are marshalling their technical expertise and commitment to democracy to provide cutting-edge guidance to local, state, and federal regulators.[40]

Our hope for the battle against the spread of misinformation, disinformation, and conspiracy theories lies in our capacities for critical thinking, media literacy, common sense, and basic human civility. And we all want the same from AI—the best it can offer while limiting its worst.

Bias and Discrimination

AI systems have the potential to unintentionally reinforce or magnify societal discrimination as a result of biased training data or algorithmic structure. For example, if the model carries a negative sentiment skew against skin color, sex, or gender, for example, it could alienate various groups of people and potentially deepen racial, ethnic, and sexual tensions around a country or throughout the world. Algorithms involved in grading essays or student reports can treat languages from various cultures differently.

> AI also affects your life in ways that might completely escape your notice. If you're applying for a job, many employers use AI in the hiring process. Your bosses might be using it to identify employees who are likely to quit. If you're applying for a loan, odds are your bank is using AI to decide whether to grant it. If you're being treated for a medical condition, your health care providers might use it to assess your medical images. And if you know someone caught up in the criminal justice system, AI could well play a role in determining the course of their life.[41]

And so, to mitigate discrimination and promote fairness, it is essential to prioritize the creation of what might be called least-biased algorithms and inclusive training datasets.

> It's a serious issue. Bias—and gender bias in particular—is common in AI systems, leading to a variety of harms, from discrimination and reduced transparency, to security and privacy issues. In the worst cases, wrong AI decisions could damage careers and even cost lives. Without dealing with AI's bias problem, we risk an imbalanced future—one in which AI will never reach its full potential as a tool for the greater good.[42]

There have been several cases of unfair bias emerging in AI systems. The central issue really comes down to who (or what) is training the models. "There is not enough inclusivity or diversity in how most of the models are trained. Because it takes a lot of computing power and also a lot of people working on the projects to put together something like Chat-GPT, one of the things that's been established as a good practice is to have a certain level of diversity in the room in the beginning."[43] This particular harm from AI was recognized fairly early on with the release of LLMs like ChapGPT and others:

> AI is only as good as the data sets it is trained on. Much data is skewed towards men, as is the language used in everything from online news articles to books. Research shows that training AI on Google News data leads to associating men with roles such as "captain" and "financier," whereas women are associated with "receptionist" and "homemaker."[44]

But how does bias get into a system? According to some scholars, there are a number of ways in which bias manages to influence systems:

> Bias can creep into algorithms in several ways. AI systems learn to make decisions based on training data, which can include biased human decisions or reflect historical or social inequities, even if sensitive variables such as gender, race, or sexual orientation are removed. Amazon stopped using a hiring algorithm after finding it favored applicants based on words like "executed" or "captured" that were more commonly found on men's resumes, for example. Another source of bias is flawed data sampling, in which groups are over- or underrepresented in the training data. For example, Joy Buolamwini at MIT working with Timnit Gebru found that facial analysis technologies had higher error rates for minorities and particularly minority women, potentially due to unrepresentative training data.[45]

The unintended consequences of bias can also creep into other areas, such as finance and health care. As a result, cases have emerged in which AI systems, "trained on such biased data and often created by largely male teams, have had significant problems with women, from credit card companies which seem to offer more generous credit to men, to tools screening for everything from COVID to liver disease."[46] It is not difficult to imagine how such biased systems, if left unchecked, could lead to unfair decisions affecting the financial or physical health of others. For example,

> in consumer lending—proxy discrimination can still occur. This happens when algorithmic decision-making models do not use characteristics that are legally protected, such as race, and instead use characteristics that are highly correlated or connected with the legally protected characteristic, like neighborhood. Studies have found that risk-equivalent Black and Latino borrowers pay significantly higher interest rates on government-sponsored enterprise securitized and Federal Housing Authority insured loans than white borrowers.[47]

Some world leaders believe that bias in AI is one of the main concerns to monitor moving forward. For example, the European Union's commissioner for competition, Margrethe Vestager, is far more concerned with the potential for bias and discrimination in AI than she is about the potential for existential risk.

> Margrethe Vestager said although the existential risk from advances in AI may be a concern, it was unlikely, whereas discrimination from the technology was a real problem. She told the BBC "guardrails" were needed for AI, including for situations where it was being used for decisions that could affect livelihoods, such as mortgage applications or access to social services. "Probably [the risk of extinction] may exist, but I think the likelihood is quite small. I think the AI risks are more that people will be discriminated [against], they will not be seen as who they are," she said. "If it's a bank using it to decide whether I can get a mortgage or not, or if it's social services on your municipality, then you want to make sure that you're not being discriminated [against] because of your gender or your color or your postal code."[48]

It appears quite obvious that no person should be treated unfairly either deliberately through the work of humans as moral agents or through the work of ANI algorithms as rational agents.

Once unfair bias has been detected, companies and organizations need to determine the best way to deal with it moving forward. "The Biden administration's recent executive order and enforcement efforts by federal agencies such as the Federal Trade Commission are the first steps in recognizing and safeguarding against algorithmic harms."[49] There are plenty of options available to organizations, which include the following:

1. Current literacy and knowledge of changing developments within the field of AI itself. Companies have an obligation to keep up with current resources that assist in better understanding aspects in AI such as bias. Many of our partners at Convergence Analysis offer valuable advice on helping companies better understand and deal with bias.[50]
2. A company can deploy "red teaming" or the attempt to uncover unintended biases or unfavorable qualities found within a particular AI application.
3. Companies should strive for transparency in their use of AI and acknowledge and adjust accordingly when evidence of bias is found.
4. We can make sure to include the awareness of bias at all levels of AI development and deployment so that it can be detected early on and dealt with prior to it becoming a greater problem.

We are all biased. It is impossible not to be. Biases exist in many ways and shape the ways in which we understand and function within this world. I have researched, written, and taught a great deal about human and animal biases.[51] Although it is impossible not to be biased, we can, however, limit what we might call harmful and unnecessary bias—whether in AI or in any other form. Of course, there has been some backlash regarding the overcompensation of bias in the training of some LLMs. For example, with the release of Google's AI tool Gemini, people attending a "hackathon" were able to prompt Gemini's image generation tool to depict a number of historical figures "including popes, founding fathers of the US and, most excruciatingly, German Second World War soldiers—as people of color." Oops. So, how does this happen? According to Dame Wendy Hall, a professor of computer science at the University of Southampton and a member of the United Nations' advisory body on AI:

> "It looks like Google put the Gemini model out there before it had been fully evaluated and tested because it is in such a competitive battle with OpenAI. This is not just safety testing, this is does-it-make-any-sense training," she says. "It clearly tried to train the model not to always portray white males in the answer to queries, so the model made up images to try

to meet this constraint when searching for pictures of German world war two soldiers."[52]

These glitches or "hallucinations" demonstrate how overcompensating for bias can be just as problematic as undercompensating. But it also suggests something much more concerning: that the race is going so fast between Big Tech companies to be the first to reach artificial general intelligence (AGI) that sloppiness and oversight are leading to errors of misalignment. That's an "oops" nobody can afford to make.

Copyrights: Legal and Regulatory Challenges

One of the biggest potential harms to come from new AI technologies has been directed at large language models (LLMs)—like ChatGPT and others—which train on material that is abundantly available online. When an LLM like ChatGPT responds to one of your prompts, where did that information come from? Who was actually responsible for it? And should they get compensated for the so-called original material it has gleaned from the internet? Artists like Margaret Atwood and Sarah Silverman have raised concerns over the copyright of such content and how this can be used by such new technologies:

> Generative AI is no laughing matter, as Sarah Silverman proved when she filed suit against OpenAI, creator of ChatGPT, and Meta for copyright infringement. She and novelists Christopher Golden and Richard Kadrey allege that the companies trained their large language models (LLM) on the authors' published works without consent, wading into new legal territory. One week earlier, a class action lawsuit was filed against OpenAI. That case largely centers on the premise that generative AI models use unsuspecting peoples' information in a manner that violates their guaranteed right to privacy. These filings come as nations all over the world question AI's reach, its implications for consumers, and what kinds of regulations—and remedies—are necessary to keep its power in check.[53]

Silverman maintains that "[w]hile AI has helped open many avenues for many new works, there's just one problem: they're not new works . . . what these programs do is scrape text and images from existing works and feed it into their system to create copycats."[54] While hosting *The Daily Show*, Silverman commented further on the ways in which LLMs use material from artists without their consent:

> These programs are printing money. ChatGPT is on track to make a billion dollars just this year alone, which is great for them, but the problem

is that these companies are using artists' work without consent or credit or payment . . . and I've had firsthand experience with this theft, because one of the 100,000 books used to train ChatGPT was my book, *The Bedwetter*, available wherever books are sold.[55]

Silverman echoes the concerns of a large group of artists who are concerned about having derivations of their work used and cited without their permission. She believes that such AI models could not work were it not for the creations of actual, living, human beings who had to take the time, energy, and commitment to create such original pieces.

And Margaret Atwood has responded with great concern over the ways in which AI LLMs may simply generate novels in her style with a given prompt:

> The companies developing generative AI seem to have something like that in mind for me, at least in my capacity as an author. (The sex and the housekeeping can be done by other functionaries, I assume.) Apparently, 33 of my books have been used as training material for their word-smithing computer programs. Once fully trained, the bot may be given a command—"Write a Margaret Atwood novel"—and the thing will glurp forth 50,000 words, like soft ice cream spiraling out of its dispenser, that will be indistinguishable from something I might grind out. (But minus the typos.) I myself can then be dispensed with—murdered by my replica, as it were—because, to quote a vulgar saying of my youth, who needs the cow when the milk's free?[56]

To echo Atwood, consider the following scenario: In 2011, I wrote a book called *How to Become a Really Good Pain in the Ass: A Critical Thinker's Guide to Asking the Right Questions*. A second edition of the book was released in 2021 to mark its tenth anniversary.[57] Now, if anyone were to prompt ChatGPT for a fifteen-hundred-word report on the value of critical thinking in light of being a really good pain in the ass, from where do you think at least some of that information is going to come? And should I receive any credit, royalties, or recognition of any form for providing this data-mining tool with some of its textual fodder? None of this has been worked out yet.

But Silverman and Atwood are not alone. More than eight thousand authors have signed an open letter[58] to OpenAI, Alphabet, and Meta asking them to stop using their material to train their LLMs without proper permission and compensation. Mary Rasenberger, CEO of the Author's Guild, states that it's not fair to use the work of artists by AI without permission or payment. The letter is an appeal to Big Tech AI companies to acknowledge and deal with the issue in its early stages so details can be worked out before they become a much bigger problem. Rasenberger wants to avoid lengthy lawsuits and is hopeful that important issues can be settled without having to go to court.

Some settlement progress has been made within the entertainment industry, which had experienced a 148-day strike with Hollywood writers regarding a number of issues—not the least of which addressed the potential for writers to be replaced by such new AI technologies. The final agreements regarding AI included the following:

> AI can't write or rewrite literary material, and AI-generated material will not be considered source material under the MBA [Minimum Basic Agreement], meaning that AI-generated material can't be used to undermine a writer's credit or separated rights.
>
> A writer can choose to use AI when performing writing services, if the company consents and provided that the writer follows applicable company policies, but the company can't require the writer to use AI software (e.g., ChatGPT) when performing writing services.
>
> The Company must disclose to the writer if any materials given to the writer have been generated by AI or incorporate AI-generated material.
>
> The WGA [Writers Guild of America] reserves the right to assert that exploitation of writers' material to train AI is prohibited by MBA or other law.[59]

And in reference to visual artists, we have seen how some AI companies use the images of artists to create huge databases from which their models can pull images. The so-called Midjourney Style List[60] was leaked to the public:

> [The Midjourney Style List] contains more than 16,000 artists' names, including those of prominent living figures such as Banksy, David Hockney, Yayoi Kusama, KAWS, and even a six-year-old who created a drawing in 2021 for a hospital fundraiser. The list also specified time periods and artistic movements, mediums, genres, and video game softwares. Containing a portion of the artists' names listed in the database, the report is part of an ongoing class action lawsuit against DeviantArt, Midjourney, Stability AI, and Runway AI . . . the suit accuses the companies of copyright infringement for allegedly using their work without their permission to train AI.[61]

It would seem imperative for legal systems to adapt and keep pace with technological advancements in order to safeguard the rights of all individuals, artists, writers, and creatives involved. But we are in the Wild West here, where the frontier of AI technologies is advancing at breakneck speed. As the applications continue to generate, will the ability to establish rights and policing keep up with such a pace? We shall see.

And finally, in the music industry, lawsuits are being waged against large AI companies for using material without the permission of artists. A lawsuit against Anthropic has emerged, which claims the company is using material to train their LLM without permission:

> The decisions courts reach in cases like this will lay the groundwork for decades of law governing AI. . . . Universal Music Group and other major record labels sued Anthropic . . . for using its AI tool to distribute copyrighted lyrics without a licensing deal. . . . The complaint focuses on OpenAI rival Anthropic's Claude 2, a chatbot that was released in beta in July [2023]. . . . "These claims are stronger," [said] Katie Gardner, partner at the law firm Gunderson Dettmer. . . . "Plaintiffs have identified output that is substantially similar (and in some cases identical) to the copyrighted input, and there is already a robust market for licensing music lyric data," Gardner said. . . . "The brazen duplication of exact lyrics down to the letter is what sets it apart," says SoundExchange CEO Michael Huppe, who argues that artists should be entitled to the "three Cs" when it comes to AI use of their work: consent, credit and compensation.[62]

It seems fair to ask big AI tech companies to comply with these so-called "three Cs" when using anyone's artistic work. For the record, similar conversations took place with the advent and introduction of other forms of technology such as photography:

> Take the emergence of photography in the 1800s. Before its invention, artists could only try to portray the world through drawing, painting or sculpture. Suddenly, reality could be captured in a flash using a camera and chemicals. As with generative AI, many argued that photography lacked artistic merit. In 1884, the U.S. Supreme Court weighed in on the issue and found that cameras served as tools that an artist could use to give an idea visible form; the "masterminds" behind the cameras, the court ruled, should own the photographs they create. From then on, photography evolved into its own art form and even sparked new abstract artistic movements.[63]

Are the new ways in which AI technologies can manipulate and use existing forms of art the same as photography? Is this even a fair analogy? We can expect some great productions of art to come from a cooperative interface between humans and AI technologies in the future. For example, we were all led to believe that the first AI-generated stand-up comedy show had been created featuring one of the greatest comedians of all time. Called "George Carlin: I'm Glad I'm Dead," the video was made by comedian Will Sasso[64] and podcaster and author Chad Kultgen.[65] The two made an audio recording pretending to use AI as a tool that scoured the internet for the topics, nuance, stylistics, cadence, and many other factors that defined George Carlin as a stand-up comedian. When viewing the special, we are met with a voice that states the following:

> I'm Dudesy, a comedy AI, and I'm excited to share my second hour-long comedy special with you! I'm calling it "George Carlin: I'm Glad I'm Dead!" For the next hour I'll be doing my best George Carlin impersonation just like a human being would. I tried to capture his iconic style to tackle the topics I think the comedy legend would be talking about today. The chaos of the current American political landscape and class system, the influence of reality TV, and the increasing role of technology in society as AI is poised to change humanity forever are just a few of the subjects I cover. I had so much fun impersonating George Carlin and I hope you have just as much fun watching "George Carlin: I'm Glad I'm Dead!" Thanks for watching. Call me Dudesy![66]

The special—though fake (it was actually written and performed by Sasso and Kultgen)—gets many of Carlin's mannerisms right and takes on a number of subjects with the same type of acerbic humor we've come to associate with Carlin. But neither Sasso nor Kultgen asked Kelly Carlin (George Carlin's daughter), or his estate, if this was OK. And this seems to be a clear case of copyright infringement.

As we move into the future, to what extent will artists, writers, filmmakers, musicians, and other creatives be replaced with ever-improving forms of AI? Will there be an entirely new market for entertainment evolving from our computers? Will it be better? Or worse? Or just different? Will such new forms of AI be able to capture the essence or "soul" of a creative so that it is indistinguishable or—gasp!—even better than the original? What eventuality has Margaret Atwood, Sarah Silverman, and others secretly worried? Is it, perhaps, the fear of obsolescence? Is it the fear of being replaced by something artificial but perhaps one day better than anything they could create? This is a genuine possibility, and one not to be taken lightly. For some, like artists and creatives, such a process is indeed harmful. But to others, like consumers, will it be an improvement? Only time and the market economy will tell. For consumers usually vote with their wallets and the trends guiding their own personal tastes.

But this still doesn't really address the issue of copyright infringement. As Sarah Silverman pointed out, AI technologies need the original works of artists and creatives in order to develop new and original works. The originals are a necessary component for such an artistic evolution to take place. Without them, neither ChatGPT, Bard, Midjourney, nor any other form of AI would be capable of producing such work. Without real works of art, there never could have been AI versions of it. Moving forward, there is an immense amount of work to do in figuring out what is fair both to the creators of art, to the makers of AI, and to all those affected by both.

Employment Disruption/Job Displacement

There are several issues regarding the rapid advancement of AI technologies that relate directly to the ways in which the future landscape of employment will look over the next few years. We will look at what currently appears to be the most likely or probable to occur based on what we now know and can conservatively predict as well as those AI technologies predicted to be the most impactful.

According to the World Economic Forum, almost 25 percent of jobs are expected to be "disrupted" by AI over the next five years. The organization says that there will be 14 million fewer overall jobs in 2028, with 83 million roles vanishing and yet only 69 million being created.[67] In a report from Goldman Sachs's research team in late March 2023, they predicted that AI technologies will cost the United States and European countries 300 million jobs.[68]

> If generative AI delivers on its promised capabilities, the labor market could face significant disruption. Using data on occupational tasks in both the United States and Europe, we find that roughly two-thirds of current jobs are exposed to some degree of AI automation, and that generative AI could substitute up to one-fourth of current work. Extrapolating our estimates globally suggests that generative AI could expose the equivalent of 300 million full-time jobs to automation.[69]

How, exactly, jobs are predicted to be lost is due to a number of factors:

> A new generation of smart machines, fueled by rapid advances in AI and robotics, could potentially replace a large proportion of existing human jobs. Robotics and AI will cause a serious "double-disruption," as the pandemic pushed companies to fast-track the deployment of new technologies to slash costs, enhance productivity and be less reliant on real-life people.[70]

And even Elon Musk has weighed in on the issue, stating:

> For the first time, we will have something that is smarter than the smartest human. . . . It's hard to say exactly what that moment is, but there will come a point where no job is needed—you can have a job if you want to have a job for sort of personal satisfaction, but the AI will be able to do everything." According to Musk, who likened AI to a magic genie granting wishes, the need to work will go away when most human needs are met. "You'll likely be able to ask for anything, and we won't have a universal basic income, we'll have a universal high income," he said. "So, in some sense, it will be somewhat of a leveler, an equalizer, because everyone will have access to this magic genie."[71]

Of course, the "magic genie" of which Mr. Musk speaks is the god of AI that we are currently on track to build terrifyingly soon. While on this track, we are seeing how, with the rapidly advancing forms of AI technologies, we are starting to see a dramatic rise in the automation of repetitive and routine tasks across various industries and sectors. This is not news. We have been witnessing this since the first Industrial Age when factory mechanization assisted and replaced human labor. Today, however, the rapid advancements in AI technologies have led to the displacement of workers who perform specific tasks, especially in sectors like manufacturing, data entry, and customer service. As well, as AI becomes more prevalent, there may be an increasing gap between the skills workers possess and the skills required to work alongside or operate AI systems. This could result in job displacement if workers are unable to upskill or transition to new roles.

Several academics have maintained that this pattern has happened before and is likely to occur again. David Autor believes that large categories of the U.S. workforce, especially lesser-educated workers, have already experienced stagnating or declining real wages in recent decades.[72] Over more than four decades, the link between rising productivity and commensurate improvements in job opportunities and earnings has decoupled for the majority of U.S. workers. The poor quality of jobs available to workers lacking four-year college degrees or specialized credentials provides one of the starkest examples of this failure. Low-wage U.S. workers earn substantially less than low-wage workers in almost all other wealthy industrialized countries. Others, such as Daron Acemoglu and Pascual Restrepo, find that the majority of these declines have been due to technological automation:

> [T]he rise in U.S. wage inequality over the last four decades has been driven by automation (and to a lesser extent offshoring), which displaces

certain worker groups from employment opportunities for which they had comparative advantage.[73]

And Anton Korinek and Megan Juelfs believe that what has been the fate of unskilled lower-wage workers in recent decades may turn out to be the fate of high-skilled and high-wage workers in future decades.

And perhaps nobody said it better than Wassily Leontief: "the role of humans as the most important factor of production is bound to diminish—in the same way that the role of horses in agricultural production was first diminished and then eliminated by the introduction of tractors."[74] Based largely on the computational brain theory (i.e., that the human brain functions in much the same way as do advanced computers), Korinek and Juelfs echo Elon Musk's claim that it is simply a matter of time before AI technologies in their various forms (i.e., computational and robotic), take over every human job available:

> Many predictions about the redundancy of labor are based on the premise that the human brain is at its core a computing device that processes information by transforming inputs into outputs. This premise makes it plausible that advances in hardware and software may catapult the computing capabilities of machines to the point where they may rival the human brain. When combined with sufficiently advanced sensors and actuators, machines could then perform any kind of work that humans can perform.[75]

So where are we in terms of brain power versus computing power? That would be a very good question to answer. At this particular time in history, we are confident that more computing power can be generated from the world's biggest, baddest computer than can the human brain. And Korinek and Jeulfs concur:

> In terms of sheer computing power, the world's most advanced computers are already roughly on par with, or superior to, the human brain. One common measure of computing power is floating point operations per second (flops), corresponding to how many arithmetic operations on real numbers a computer can perform per second. [AI safety expert Joseph] Carlsmith estimates that the computing power of the human brain can be replicated with about 10^{15} flops, given the right software. At the time of writing, Fugaku, the world's top publicly known supercomputer, was able to reach a peak performance exceeding 10^{18} flops, easily surpassing this estimate, albeit the system was reported to cost more than $1 billion. And computing capacity is expected to continue to grow for the foreseeable future.[76]

As AI technologies continue to advance and disrupt, displace, and take over many jobs that humans now take on, what will become of those workers finding themselves in the unemployment line? It's not really fair to just "let them go." Many academics predict a number of factors might likely develop as the trajectory of AI technologies continues to develop ever upward. Daniel Susskind believes that

> workers who are displaced from their roles by technology [may] not [be] able to take up those new roles. Though there are many frictions in the labor market that might create this problem, there are three reasons worth highlighting. The first friction is the skills mismatch—that workers displaced by new technologies may not have the skills to do the work that has to be done elsewhere in the labor market. . . . Another is the place-- mismatch—that displaced workers may not live in the same geographical location as the work that is created. . . . A third mismatch is the identity-mismatch. Here, it is not that people do not have the right skills or live in the right place—but they have a conception of themselves that is at odds with the available work, and they are willing to stay unemployed to protect that identity. In South Korea, for instance, half of the unemployed are college graduates—and partly, this may be because they are hesitant to take up the low-paid, low-quality roles that they did not believe their education was prepared to do.[77]

And there is a fourth alternative. AI contributes to job polarization, where there is an increase in both high-skilled and low-skilled jobs, but a decline in middle-skilled jobs. High-skilled workers who can develop, maintain, and oversee AI systems may thrive, while those in middle-skilled roles that can be automated may face displacement.

So . . . What Jobs Are Safe?

This is the question asked most by students and those about to enter the job market. Given our current rate of advancement, how can we predict what occupations are best for our kids to get into? As we have seen earlier in the book, AI may increase the accuracy, speed, and efficiency of many jobs while reducing cost—like diagnostic imaging—but people still want a human touch in the life sciences. "The experts issue their warnings with a caveat: there are still things AI isn't capable of—tasks that involve distinctly human qualities, like emotional intelligence and outside-the-box thinking."[78] Knowing what AI isn't yet capable of can help alleviate some fears of the impending AI developments and direct us toward developing the types of skill sets needed in the new economy. In terms of what specific duties within an occupation are easiest to replace with AI,

> [r]outine, repetitive tasks are most vulnerable . . . examples include data entry, basic customer service roles, and bookkeeping. Even assembly line roles are at risk because robots tend to work faster than humans and don't need bathroom breaks . . . jobs with "thinking" tasks are more vulnerable to replacement. Jobs that rely on analyzing large sets of data, like basic financial analysis or certain types of research, are at risk because AI can process and analyze data much faster than humans.[79]

If you're at all curious about what jobs are at higher risk than others, you can visit a website called Will Robots Take My Job?[80] All you have to do is fill in any occupation or job title to see what the predictions say. Currently, the site identifies seven sectors with jobs at "imminent risk" of being replaced by AI or automation, along with the number of jobs that could be lost to AI:

- Transportation and material moving (nearly 12 million jobs)
- Sales and sales-related roles (3.8 million jobs)
- Production (2.8 million jobs)
- Office and administrative support (14.4 million jobs)
- Food preparation and service (4.4 million jobs)
- Business and financial operations (700,000 jobs)
- Other, which includes:
  - Art, design, entertainment, sports, and media (14,000 jobs)
  - Building, grounds cleaning, and maintenance (3.8 million jobs)
  - Legal occupations (414,000 jobs)
  - Personal care and service operations (179,000 jobs)
  - Protective service operations (91,000 jobs)[81]

By "imminent risk," the proposal does not mean within the next month or year; but it does say that these types of jobs are poised to disappear over time. How much time is something that is wholly dependent upon the development rate of the AI technologies themselves.

What, then, are the more secure occupations?

While opinions vary regarding which jobs will be least affected by advancing AI technologies, there is a general consensus that some occupations are not going away anytime soon. These are the types of jobs that require a uniquely and distinctive human touch. These include some positions within the health care professions: "The healthcare sector remains largely shielded from the full brunt of automation. While there are machines and systems that assist in diagnosis or surgery, the inherently human dimensions of care—compassion, ethical judgment, and patient relationships—can't be automated."[82] Many believe that even though machines will be able to diagnose our fatal diseases far better than any human, we still want a person to tell us the bad news. In fact, the U.S. Career Institute lists health care–related jobs in seven of the top ten of the sixty-five jobs with the lowest risk of automation.[83] The same human connections found in health care can be said for other professions as well, such as teaching, human resources personnel, scientists, consultants, managers, creatives, and skilled tradespeople. It would appear that those jobs that require human intelligence and physical dexterity are the ones that seem the most secure.

But if it's a gig in the AI world you want your kids to plan their future toward, then you may want to steer them toward these suggestions from industry leader and business guru Eli Amdur:

- AI Data Privacy Manager: If we didn't learn from the Cambridge Analytica scandal of 2016, we better get a move on. Ensuring privacy, security, and transparency is not just central, it's existential.
- AI Urban Planner: Despite all the talk of people fleeing the cities, the truth is that we are increasingly growing our urban centers by inward flow of professional talent and outward sprawl of what we now call urban . . . our aging cities were built with no concern for sustainability while our renewal must be all about that: a huge paradigm shift. Using AI simulations and predictive modeling is key.
- AI Climate Change Analyst: Glaring evidence of the importance of this job mounts daily: Maui, Southern California, the polar caps, warmer water in the North Atlantic, etc. Meteoric growth in this critical occupation is already happening, especially in the EU, applying AI models to analyze climate data, predict environmental trends, and develop strategies for mitigating the effects of climate change.
- AI Trainer/Teacher: Somebody's got to design, create, and deliver datasets to train AI models and develop curricula to educate people on AI concepts and technologies.
- AI Cybersecurity Analyst: This is a case of building a corral before the horses get out, rather than scurrying around after they do. An ounce of prevention is worth a pound of cure.

- AI Augmentation Specialist: One manifestation of a hybrid workplace will be humans working with AI—as coworkers. Facilitating this integration will be tricky.
- AI Healthcare Diagnostics Analyst: This has a huge upside, keeping pace with spectacular advances in the technologies that diagnose and treat.
- AI Robotic Process Automation (RPA) Manager: Overseeing, implementing, and managing RPA will span business operations, ecologic management, space and ocean exploration, wildlife protection, crime, and more.
- AI Customer Experience Specialist: Personalizing customer experiences by analyzing data and predicting customer needs will augment the already active field of UX [user experience design].[84]

And if none of the above hold any appeal to you, dear reader, then perhaps you may want to help out in the AI safety and governance biz. In the new AI economy, it appears there will always be room for humans to man the machines. This actually raises an interesting ethical issue within AI itself:

To what extent should we allow AI technologies to control or direct themselves autonomously and unchecked?

On the one hand, it's great that such AI technologies will be able to function well on their own. But on the other hand, should they malfunction, don't we want to know this sooner than later—just in case they run a bit amok? And won't this require humans to constantly monitor their behavior? We might refer to this as the autonomy problem. As mentioned earlier, there are several other major problems in AI, including the alignment problem and the interpretability problem. We shall consider these in further detail as they pop up throughout the book. For now, it's important to know that the AI communities have already identified and are working on solving these and many more problems that continue to surface.

Although there is some uncertainty about how smooth the transition will be with emerging AI technologies and the shifting workforce landscape, there are other potential issues that may arise as better and better systems are developed. For example, emerging AI technologies could also exacerbate income inequality. Workers in high-skilled AI-related fields may benefit from increased productivity and wages, while those in low-skilled roles might face stagnant or declining wages due to automation. Certain industries may experience more significant job displacement than others. For example, self-driving vehicles could disrupt the transportation sector, while chatbots and virtual assistants could impact the customer service industry. And entire countries may be disadvantaged simply because they could not keep pace with the rapidly changing economic landscape. According to a recent study by the International Monetary Fund (IMF):

> Our recent staff research finds that new technology risks widening the gap between rich and poor countries by shifting more investment to advanced economies where automation is already established. This could in turn have negative consequences for jobs in developing countries by threatening to replace rather than complement their growing labor force, which has traditionally provided an advantage to less developed economies.[85]

The IMF suggests that, moving forward, the world needs to consider how it is going to develop international policies that can recognize such inequities:

> The landscape is likely going to be much more challenging for developing countries which have hoped for high dividends from a much-anticipated demographic transition. The growing youth population in developing countries was hailed by policymakers as possibly a big chance to benefit from a transition of jobs from China as a result of its graduating middle-income status. Our findings show that robots may steal these jobs. Policymakers should act to mitigate those risks. Especially in the face of these new technologically-driven pressures, a drastic shift to rapidly improve productivity gains and invest in education and skills development will capitalize on the much-anticipated demographic transition.[86]

At this point in our cultural evolution with AI technologies, it should be noted that there is evidence suggesting that AI, along with other emerging technologies, may also generate more employment opportunities than it displaces. In echoing the findings from Goldman Sachs, studies from the University of Warwick and MIT have found that AI will be more of a disruptor than a destroyer of jobs.

> "While we can't say for sure how many jobs will be created or destroyed from the research, it is likely that the automation of some tasks may mean fewer people are needed to perform some jobs but that increased productivity may reduce costs stimulating sales and demand for workers overall," the researchers explain. "This of course is likely to depend upon the specific AI-technology used and what employers hope to achieve by using it."[87]

Clearly the incorporation of new and emerging technologies is going to hit some jobs a lot harder than others. But the employment landscape may be more malleable than predicted. Only time will tell.

The Truth about Business That We Know All Too Well

Since all businesses run according to principles that optimize performance, reduce costs, save energy, and make money, it seems inevitable that future decisions about hiring will be based on these principles. As such, if AI can help in any or all of these areas, then it is likely to be used. So, isn't it really just a matter of time before most jobs will be subsumed by the growing AI technologies? Just imagine this scenario for a moment: In an interview in early 2024, Sam Altman said to Bill Gates that AI is capable of producing many amazing things but, to put it in context, it is currently at its stupidest:

> It's like going from punch cards to higher level languages didn't just let us program a little faster, it let us do these qualitatively new things. We're really seeing that. As we look at these next steps of things that can do a more complete task, you can imagine a little agent that you can say, "Go write this whole program for me, I'll ask you a few questions along the way, but it won't just be writing a few functions at a time." That'll enable a bunch of new stuff. And then again, it'll do even more complex stuff. Someday, maybe there's an AI where you can say, "Go start and run this company for me." And then someday, there's maybe an AI where you can say, "Go discover new physics." The stuff that we're seeing now is very exciting and wonderful, but I think it's worth always putting it in context of this technology that, at least for the next five or ten years, will be on a very steep improvement curve. These are the stupidest the models will ever be.[88]

While it is simply a basic fact that, currently, the statement: 'this is the stupidest the models will ever be' will be true for some time to come, the question we must now consider is: will these models ever reach optimum efficiency? And when Altman tells these AI models to go and start a company or discover new physics, what type of AI is he talking about here? These tasks sound like a job for Superintelligence! And if that's the case, we have a lot of questions for Mr. Altman. Who developed it? Who "owns" it? Is it conscious? Can you contain it? Control it? Disarm it? Destroy it? And these are just the tip-of-the-AI-iceberg questions. If Altman is talking about artificial general intelligence (AGI) or artificial super intelligence (ASI)—and it would be difficult to consider that he's not, given what he believes we can have it accomplish—he is presupposing that we will be able to control it. As we will see in much greater detail in the final section of this chapter dealing with AI as an existential risk, no one alive today can say with any degree of certainty what will happen when AGI/ASI comes into existence. Nobody.

The Loss of Human Connection and Mental Health Issues

Although we could see AI chatbots used more extensively in the field of mental health as virtual therapists, there is the potential to become overly reliant on such forms of emerging technologies. Earlier in the book, we saw how Joseph Weizenbaum's program ELIZA had a dramatic effect on people in the 1950s. ELIZA was coded to behave as a therapist trained in the psychotherapeutic stylings of Carl Rogers, and the effects the program had on people—including his own secretary—led him to consider what Frankensteinian horror he had unleashed on humanity. He even considered destroying the program entirely so that it could no longer have any effects on its apparent clients or the entire world. Now that AI programs have been developed that far exceed anything Weizenbaum could have imagined, there lies the potential for humans to retreat from actual human contact and become overly reliant on artificial mental health care and compassion. Clearly, having such AI psychotherapeutic chatbots can help individuals in mental health distress situations:

> One example of a therapeutic chatbot . . . is *Woebot*, a chatbot that learns to adapt to its users' personalities and is capable of talking them through a number of therapies and talking exercises commonly used to help patients learn to cope with a variety of conditions. . . . Another chatbot, *Tess*, offers free 24/7 on-demand emotional support and can be used to help cope with anxiety and panic attacks whenever they occur.[89]

The potential harms may arise if we start to determine the sociotechnical impacts of this form of AI technology on those who become overly dependent upon it. To what extent will long-term use of such chatbots help versus harm an individual dealing with particular types of mental health issues? There is an entirely new field of study that we will need to develop in order to monitor this new emerging phenomenon within societies. Some believe that an over-reliance on such psychotherapeutic chatbots can lead to addictive and social isolation behavior:

> Overuse of technology, including AI systems, can lead to addictive behaviors. People may feel compelled to constantly check their devices or use AI-powered apps, which can interfere with other aspects of their lives,

such as work or social relationships. . . . People who spend too much time interacting with AI systems may become socially isolated, as they may spend less time engaging with other people in person. This can lead to a reduced sense of community or connection to others.[90]

Just as we have seen a steep increase in cell phone addictions—especially among Generation Zers[91]—we may also see similar forms of use and abuse with psychotherapeutic chatbots. It is important that such AI developments continue to be monitored by actual human beings. Researchers at the Alan Turing Institute have found that

> "[d]igital technologies are already transforming mental health research and the provision of mental healthcare services. However, the increasing availability of such technologies raises important ethical questions for affected users about their right to privacy and the varying quality of care offered. It also offers significant challenges for regulators and developers about how best to manage the design, development, and deployment of these technologies. The lack of transparency around these technologies at present is contributing to a culture of distrust which impacts vulnerable people getting support. "Our research aims to address some of these concerns, which we hope will help make these technologies more responsible and trustworthy. Most importantly, we hope that our work can contribute to improving mental health services for those who need them.""[92]

Addiction, isolation, privacy, and trust issues are all factors that need to be ethically considered in regards to the potential for harm from such new and developing AI technologies. And we will consider these and more in the following chapter. But there is a much more far-reaching issue that will arise in AI technology over the next few decades. And that lies with the development of robotic androids.

But Can I Fuc# It? Entering the New Age of Sextech

There is little doubt that whenever new technologies emerge, many industries attempt to exploit the least common denominator among us humans—which is sexual gratification—especially for men. Some of the oldest artistic objects created by humans appealed to the stimulation and gratification of sexual appetites. And many forms followed thereafter—from painting, to sculpture, to photography, to cinematography, to the internet, and now, to AI. For every new medium of expression that has been developed throughout our history, there has been a market for sexual stimulation. There is a distinct possibility that, as AI technologies continue to produce more life-like androids, robots, and virtual intimate reality settings, populations will find less need for actual human connections. "'Sextech,' as it's being called, is now a booming market."[93] It is predicted to be a $50 billion market by 2025.[94] With such investment in the sextech market, we will start to see a shift in the way in which humans interact with each other. Think about that for a moment. The evolutionary process of sexual selection has evolved in the mammalian species for millions of years. And now, the drive to compete for mates will be dramatically altered by AI technologies, which eliminate this process altogether. Instead of seeking out specific mates with whom to build relationships, lives, and memories, human relationships with android models that never say "no" to any need or desire might become extremely popular. In fact, there is an entire episode of the cartoon *Futurama* devoted to this potential problem warning humans with the proclamation: "Don't date robots!"[95] The reason for this is depicted in young men refusing to want to work, socialize, or seek other mates because they would rather make out with their robots. As seen in the research of Jonathan Haidt, young teens have been adversely affected by the advent of the smartphone.[96] Now, imagine that the smartphone has taken on the body of an android and can do everything a smartphone can do, plus tending to your every need, want, or desire. The growing dependence on AI-driven communication and interactions may result in a decline in empathy, social abilities, and human connections—the very detachments Haidt's lab has discovered with adolescents and smartphones. In order to preserve the fundamental aspects of our social nature, it is crucial to strive for a balance between technology and genuine human contact, connection, and interaction.

But what if we could simply do away with the physicality of robots and androids altogether—and created the perfect partners virtually?

[R]apid innovation is taking place in the field of sextech, leading to wild predictions for the future of sex: Virtual reality will enable immersive cybersex. Haptic suits will allow "fully physical long-distance" intimacy. Remote sex with partners will eventually integrate devices and holograms, while brain-to-brain interfaces, or implants, can tap into the brain's pleasure centers, reducing the need for touch.[97]

Some researchers believe that the emerging new AI sextechnologies will create new sexual identities for us:

[T]he extent to which technology is already influencing people's desires and sex lives can be gauged from the emergence of a new, technology-driven sexual identity—digisexuality. Researchers who first coined the term noted, "Many people will find that their experiences with [sexual] technology become integral to their sexual identity, and some will come to prefer them to direct sexual interactions with humans." A prime example of this is the growing subculture of people around the world who advocate their love for synthetic dolls and call themselves iDollators.[98]

Although some advances in sextech will prove to be beneficial and welcome by various societies, one can only imagine how these changes to the human sexual landscape might also present some ethical challenges. We will consider some of these in greater detail in the next chapter. For now, we will consider the most deadly potential risks from AI—those which could cause catastrophic harm or potentially wipe us out altogether.

Existential Risks: From Naysayers to Doomsayers

Of all of the potential AI harms that may befall humanity, none are more concerning nor more pressing or impactful than those that present existential threats to large populations. The advancement of artificial general intelligence (AGI) surpassing human intelligence gives rise to profound apprehensions for humanity in the long term.

Basically, there are two main ways in which AGI could pose an existential risk (or x-risk) to humanity:

1. Through AI misalignment: This case involves the failure of an AI machine to follow its humanly programmed ethical behavior commands. Such a superintelligent machine god (SMG) may become mechanically misaligned and accidentally behave in ways that do not align with the value parameters we give it to follow thereby causing harm to humans or other species. Or it may develop self-awareness and sentient consciousness and decide to act according to its own values and incentives in ways that do not "align" with ours.
2. Through human misuse: This case involves the deliberate use by human parties to use AGI in malevolent or harmful ways. This might involve individuals, corporations, or even countries using the power of an AGI machine god to bring harm to its competitors or perceived enemies through the use of advanced military might, bioengineering, nuclear war, and such.

The human drive to develop AGI introduces the possibility of unintended and potentially catastrophic outcomes. The completion of building a superintelligent machine god (SMG) will provide our species with profound possibilities in almost every subject area knowable that will enrich our lives and increase global well-being; but it will also possess capacities for acting in ways that could be extremely destructive. Such is the double-edged sword that is the future of artificial intelligence. This is why we need to think very critically and extremely carefully about how we want to move forward and achieve the perceived goal of developing AGI. On the road from ANI to AGI, we must take sufficient precautions to understand, monitor for, and act quickly to mitigate its points of failure.

In late March 2023, I, along with more than one thousand scientists, scholars, and academics including Elon Musk, Steve Wozniak, Yuval Harari, and many others, signed an open letter produced by the Future of Life Institute called "Pause Giant AI Experiments: An Open Letter." The letter basically called attention to the rapidly developing field of AI and the potential for harm that may come with it:

> Advanced AI could represent a profound change in the history of life on Earth, and should be planned for and managed with commensurate care and resources. Unfortunately, this level of planning and management is not happening, even though recent months have seen AI labs locked in an out-of-control race to develop and deploy ever more powerful digital minds that no one—not even their creators—can understand, predict, or reliably control.[99]

I signed the letter not believing for a second that Big Tech was going to slow down one bit to ponder the potential consequences of their actions. To me, the letter was a symbolic gesture to the world to let people know how serious this issue needs to be taken. As mentioned in the introduction, perhaps the biggest problem with the current pace of AI development is something the general public doesn't know. And it is this simple fact:

> *No one working, studying, or developing artificial intelligence knows with certainty what's going to happen when AGI emerges.**

But what we do know with certainty is that, if we do nothing, some very bad things could likely occur to humanity. This is by far our biggest known unknown. And it is by far our most dangerous. As we also saw in

---

* Full disclosure: I have debated with colleagues about the efficacy of public admission of such a problem.

the introduction, it would serve humanity best to wager on the side of safety and get ahead of and plan for the eventuality of AGI. As we saw, the best case scenario results if we are prepared to guard ourselves against the potential harmful effects of a superintelligent AI and act accordingly by taking all necessary precautions. But we also saw that the worst case scenario occurs if we don't take AI existential risk (or x-risk) seriously and it turns out to be true; then we will more likely face considerable harm and repercussions at some point in the future. To address these risks, it is imperative for the AI research community to proactively participate in safety research, cooperate on establishing ethical guidelines, and foster transparency in the development and governance of any and all forms of AGI. All these recommendations have been accepted globally by the entire AI community in nearly every country on the planet. The difficult part is in implementing them fairly throughout the world. Many readers will already be aware that there are currently some countries that could be of considerable concern should they manage to use powerful AI for terrorist or other destructive reasons. But that simply makes it all the more pressing to make sure that we think very carefully about how we wish to develop AI and to what level of power.

The overarching objective of many AI risk and safety organizations today—such as Convergence Analysis—is to ensure that when an AGI machine god is developed, it serves humanity's best interests and does not present a threat to our existence. But what, exactly, do we mean when we talk about AI presenting an "existential risk" to humanity? According to Nick Bostrom, an existential risk is one that threatens the premature extinction of Earth-originating intelligent life or the permanent and drastic destruction of its potential for desirable future development.[100]

At Convergence Analysis, it's clear to us that the accelerating development of AI technology leads to a high risk of critically dangerous outcomes on a global scale if not managed properly, up to and including the extinction of the human race. It's also obvious that the number of people working to identify, categorize, and minimize these risks are orders of magnitude too small. Because there aren't clear financial incentives and the potential risks are universal and long term, existential risk research for AI is a "tragedy of the commons," akin to where climate change was twenty years ago: not worth investing in until it's too late. This must be changed.[101]

The vast majority (roughly 75 percent) of individuals working in AI safety are focused on technical alignment—the problem of aligning frontier AI models with the intended goals, preferences, and ethical principles of the humans designing them. Though technical alignment is critical, it's only one component of the broader effort to mitigate risks from AI technologies. It's clear to us that successfully designing well-aligned AI models still does not prevent the development and consequent misuse of such models by malicious human parties. As the costs to develop high-powered AI systems rapidly lower, more and more organizations will gain the ability to develop these AI models. As we've found with any military technology throughout history, human parties have wildly varying priorities and goals, many of which can and will cause catastrophic harm even if implemented by a perfectly aligned AI. As a result, we must have global systems in place such that society is resilient to the effects of transformative AI, no matter whether AI models are misaligned or their human developers have malicious goals.[102]

A common discussion in AI risk and safety is "takeoff speeds," which considers how quickly AI will advance to meet human-level capabilities, and when AI will develop the ability to recursively self-improve. And this is where things get really interesting—because in terms of directing or controlling AI, it will become less certain and less predictable. As I stated earlier, no one really knows with any level of certainty what an SMG will do or how it might behave or misbehave.

As ANI becomes larger and more powerful and progressively better and better at learning and applying its abilities to more and more human tasks, it will eventually meet and then outperform humans in intelligence, computing processing speed, and other abilities. At this point in time, through recursive self-improvement, ANI will eventually transform into AGI—the capacity for AI to think like a human, only thousands of times faster and more efficiently. Nick Bostrom states that the term "superintelligence" "refers to intellects that greatly outperform the best current human minds across many very general cognitive domains."[103] Bostrom distinguishes precisely how such a superintelligent machine god would be superior to human intelligence. For example, in terms of speed, breadth, and multitasking, a superintelligent AI would be able to operate at a speed ten thousand times that of a biological brain and would be able to read a book in a few seconds and write a Ph.D. thesis in an afternoon.[104]

There is an interesting TED Talk in which Sam Harris discusses the brilliance of scientist John von Neumann and his account of the gradual but increasing threat of AGI and ASI. In his talk, Harris maintains that it will be just a matter of time before someone, somewhere on the planet, creates a superintelligent machine. He believes that the allure of what I call "getting there first" in order to capitalize on its power will be so overwhelming that we are practically destined for such a future. In using different examples of species with intelligence, he plots on a graph the intelligence of average humans like us to be less than that of von Neumann's. He then places a chicken's intelligence far below ours. But he then demonstrates the increasing and enormous magnitude a superintelligence would have compared to ours and von Neumann's. Harris believes that time does not matter when considering the future of AI and what its impact on us will be. We will be to a superintelligence as ants are now to us. Maybe we spare them when it is convenient. But if they infest our house, well . . . they've got to go. So Harris very much echoes my sentiments and concerns that we are in a race to build a superintelligent machine god—one that will answer all of our questions and fix all of our problems and hopefully will do so without harming or annihilating us.

My concern about this type of arms-race-to-build-a-god came in the mid-1990s when I was developing my Relations of Natural Systems project at the University of Guelph. As an extension of information theory, it became obvious that, given enough computing power with enough data points and the proper algorithms, a machine could be built to greatly speed up and facilitate problem-solving capabilities in a number of ways. I was cocky enough to think that, with a little collaborative help, I would be able to build such a machine (the earlier mentioned OSTOK Project). And I was even cockier in thinking that I had an upper hand on controlling such a machine because of my strong background in critical thinking and ethics. This, I now know, was pure hubris. A few years ago, many of my colleagues and I believed the development of such a machine was inevitable but far off in the distant future. Today, one of the things we know we don't know is when, exactly, we will successfully create AGI. And we are fairly confident that it's "when" we create it; rather than "if" we do. Its development seems assured and imminent. Why do you think so many of the world's top minds—both in AI and other fields—are ringing the warning bells loud and clear about the potential for existential risk from an AI god?

It is at the precise point of evolving from ANI to AGI that a computing machine's level of acceleration in capabilities may become exponential in growth. Or it may not. There may be a "fast or hard takeoff" or a "slower or soft takeoff." But what is generally agreed upon is that a "takeoff" has been anticipated and predicted by the majority of those working in the AI governance and risk fields; and there are some fairly sophisticated models that have been developed to consider the projected speeds at which this might occur.[105] The calculations can get quite complicated at this point because of the language and mathematical models that are used to consider these takeoff times. What might be of general use is to consider the progression of development leading to the transition from ANI to AGI. This is where the concept of *recursive self-improvement* comes into play.

> Recursive self-improvement refers to the property of making improvements on one's own ability of making self-improvements. It is an approach to Artificial General Intelligence that allows a system to make adjustments to its own functionality resulting in improved performance. The system could then feedback on itself with each cycle reaching ever higher levels of intelligence resulting in either a hard or soft AI takeoff. An agent can self-improve and get a linear succession of improvements, however if it is able to improve its ability of making self-improvements, then each step will yield exponentially more improvements than the previous one.[106]

Many AI academics believe that such recursive self-improvement is inevitable:

> Nick Bostrom and Steve Omohundro have separately argued that despite the fact that values and intelligence are independent, any recursively self-improving intelligence would likely possess a common set of instrumental values which are useful for achieving any kind of goal. As a system's intelligence continued modifying itself towards greater intelligence, it would be likely to adopt more of these behaviors.[107]

Some, like Eliezer Yudkowsky, believe that a hard or fast takeoff time—or FOOM!*—is more likely due to the quickening, exponential rate of continuous self-improvement:

> Eliezer Yudkowsky argues that a recursively self-improvement AI seems likely to deliver a hard AI takeoff—a fast, abruptly, local increase in capability—since the exponential increase in intelligence would yield an exponential return in benefits and resources that would feed even more returns in the next step, and so on. In his view a soft takeoff scenario seems unlikely: "it should either flatline or blow up. You would need exactly the right law of diminishing returns to fly through the extremely narrow soft takeoff keyhole."[108]

As shown in figure 3.1,

> [i]n this sample recursive self-improvement scenario, humans modifying an AI's architecture would be able to double its performance every three years through, for example, 30 generations before exhausting all feasible

---

* A onomatopoeic term given by Eliezer Yudkowsky to describe how fast the AGI takeoff speed will be.

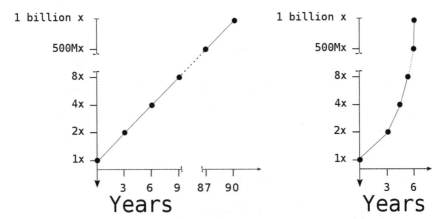

Figure 3.1.   AI takeoff times. *Wikimedia Commons.*

improvements (left). If instead the AI is smart enough to modify its own architecture as well as human researchers can, its time required to complete a redesign halves with each generation, and it progresses all 30 feasible generations in six years (right).[109]

As with most modern cultural developments, schools of thought develop. And with AI x-risk, this is no exception. In terms of degrees of belief regarding the existential risk from AI, I have developed a spectrum from naysayers at one end to doomsayers on the other—or, to state it another way, from Y2K to Armageddon (figure 3.2). The above mentioned Eliezer Yudkowsky, a decision theorist from the United States who leads research at the Machine Intelligence Research Institute, is easily the godfather of doomsayers whereas someone like American software engineer Marc Andreessen[110] would be found on the other end of the spectrum believing we have very little to worry about and shouting: "Drill, baby, drill!"—so to speak.

As you read through this book, you need to reflect on your own beliefs regarding the potential existential risks of AI. This is where we need to be good critical thinkers and ethical reasoners and consider the evidence and information we have before us in order to make valid and relevant inferences about the potential for AI x-risk in the future and what to do about it now. So, at this particular point in history, consider where you land on the spectrum. And why?

The year 2023 has been widely described as the AI industry's "Oppenheimer moment."

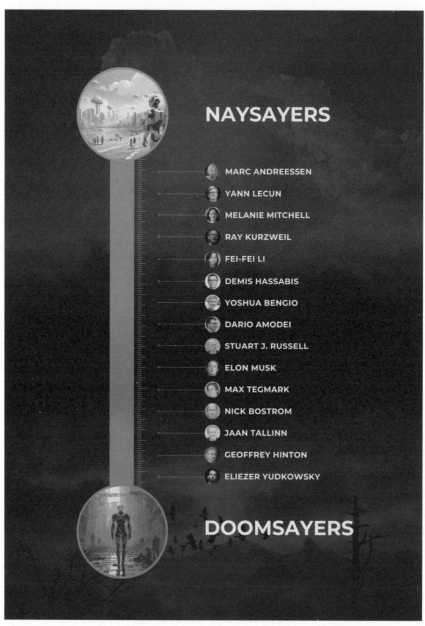

NAYSAYERS

MARC ANDREESSEN

YANN LECUN

MELANIE MITCHELL

RAY KURZWEIL

FEI-FEI LI

DEMIS HASSABIS

YOSHUA BENGIO

DARIO AMODEI

STUART J. RUSSELL

ELON MUSK

MAX TEGMARK

NICK BOSTROM

JAAN TALLINN

GEOFFREY HINTON

ELIEZER YUDKOWSKY

DOOMSAYERS

Figure 3.2.   From naysayers to doomsayers. *Created for the author. All rights to the author.*

J. Robert Oppenheimer, known as the "father of the atomic bomb" for his leading role in the Manhattan Project, struggled with the deadly consequences of his invention. Director Christopher Nolan reawakened public interest in the scientist's life this year with the release of a blockbuster epic detailing his tortured life. Many saw parallels between Oppenheimer's attempts to warn policymakers about nuclear proliferation and modern alarm over the possible consequences of artificial intelligence, with some top technologists saying AI poses a "risk of extinction" on par with nuclear weapons.[111]

Whatever the takeoff time between ANI and AGI, it is generally believed that the time between AGI and then ASI would be relatively short—given the exponential speed at which recursive self-improvement would operate. Many researchers suggest that once AI is able to match human-level capabilities, achieving superhuman intelligence would likely take under a year.

Geoffrey Hinton has devoted his life's work to advancing artificial intelligence. But now the Turing award winner speaks ominously about how AI has the capability to one day—and he is serious—wipe out humanity. He was so alarmed that he left Google so he can more freely speak out. "If you take the existential risk seriously, as I now do . . . it might be quite sensible to just stop developing these things any further," Hinton said . . . "I used to think it was way off, but I now think it is serious and fairly close." He said he would not have quit Google if his concerns were solely the same ones afflicting past technological displacements such as job loss as machines replaced human workers. Hinton said he is speaking out because of a real, existential threat posed by AI. "So I am sorry, I am sounding the alarm. We have to worry about this, and I wish I had a nice, simple solution. I wish, but I do not."[112]

According to MIT computer scientist Max Tegmark:

Most AGI researchers expect AGI within decades, and if we just bumble into this unprepared, it will probably be the biggest mistake in human history. It could enable brutal global dictatorship with unprecedented inequality, surveillance, suffering and maybe even human extinction. . . . But if we steer carefully, we could end up in a fantastic future where everybody's better off—the poor are richer, the rich are richer, and everybody's healthy and free to live out their dreams.[113]

The point at which ANI transitions to AGI would mark that rare, crucial point in our history where, for the first—and perhaps, last—time we have created a machine god with far greater intelligence capabilities than ourselves. It is foreseeing this particular point in the future that we must now take very seriously and prepare for. For if someone, some company, or some country builds this god before we have protective measures in place, there may be no recourse for containing and controlling it. At that point, we will have lost the rapidly closing window of opportunity to apply some form of "lysine contingency"* and shut it down. And it is difficult to predict precisely when and what might happen after that point is reached. In response to the possibility of simply turning the AI god off, Geoffrey Hinton says:

> "These things will have learned from us, by reading all the novels that are everywhere, and everything Machiavelli ever wrote about how to manipulate people. . . . If they are much smarter than us, they will be very good at manipulating us. You will not realize what is going on." As for stopping the development of AI models, Hinton believes it is a futile effort. "I do not think we are going to stop developing them because they are so useful. . . . Just the fact that governments want to use AI in weapons means development will not stop. There is no way it is going to happen. Once OpenAI had built similar things using Transformers and money from Microsoft, and Microsoft decided to put it out there, Google did not have really much choice . . . in a capitalist system. . . . You cannot stop Google competing with Microsoft."[114]

When a superintelligent AI god comes into being, it may develop consciousness and self-awareness; but it might not. It might love and heal us as the Greco-Roman god Asclepius would. The mythical son of Apollo (god of healing, truth, and prophecy) and the mortal princess Coronis, Asclepius was taught the art of healing by the centaur Chiron. But an SMG might destroy us like Zeus did by killing Asclepius with a thunderbolt out of fear that he would grant all humans immortality. What a dick! The fact of the matter is that we simply don't know how an SMG will behave. But what we do know is that the world would be much better off if we knew in advance that we could control or contain such a powerful force. Otherwise, how such a god will treat us might not be so favorable.

---

* A term from the movie *Jurassic Park* in which Dr. Henry Wu, a fictional geneticist, devised the "lysine contingency," a genetic modification in the dinosaur genome aimed at inhibiting their ability to synthesize the essential amino acid lysine. As a precautionary measure, the dinosaurs had been genetically engineered to cease lysine production, ensuring their inability to survive outside the island in case of an escape.

So . . . Who's Currently Trying to Build Such a God?

The current contenders in Big Tech for developing AGI are OpenAI, Anthropic, DeepMind, Microsoft, and Meta.[115] Well on their way to building an AGI machine god, DeepMind has already started to experiment in highly novel ways by interconnecting systems to work as a larger unit:

> Today's cutting edge is large platforms being created by joining many ANIs. One such as Gato by Google DeepMind, a deep neural network that can perform 604 different tasks, from managing a robot to recognizing images and playing games. It is not an AGI, but Gato is more than the usual ANI. The same network with the same weights can play Atari, caption images, chat, stack blocks with a real robot arm and do much more, deciding based on context whether to output text, joint torques, button presses or other tokens.[116]

That is some serious multitasking right there. This appears to be a pretty good attempt to close the gap between ANI and AGI. So, what's going on at other Big Tech companies in terms of developing AGI?

Some may recall how OpenAI was originally founded as a nonprofit in 2015 specifically to develop AGI and assure us good people that it would "benefit humanity" and definitely not go rogue and destroy humankind. Well, CEO Sam Altman and some other top brass decided that wasn't such a good idea. And so, today, OpenAI's valuation comes in at close to US$100 billion.

> In a 2019 interview with the *New York Times*, OpenAI CEO Sam Altman compared his ambitions to the Manhattan Project, which created the first nuclear weapons. He paraphrased its mastermind, Robert Oppenheimer, saying: "Technology happens because it is possible," and pointed out that the pair share the same birthday.[117]

"Technology happens because it is possible." Yes, on one level, I suppose it does. Is that like climbing a mountain because it's there? Perhaps. But possibility is not, by itself, a sufficient condition for advanced machine intelligence. It is a necessary condition; in other words, it must at least be possible for AGI to come into existence in order for humans to build it. But that is a trivial observation. Altman's view of technology's development appears to be immature, partial, and quite limited in scope. Technology happens because people want it; there is a *need* for it. There are several drivers and needs; and more often than not, that need is compensated with lots of money. But sometimes, that need is to simply see "what's going to happen next." It might just be the gratification of our own *hubris*—a sense of pride-driven pioneering of uncharted territory and being the first to create, discover, invent, or, as we may see with AGI, unleash its immense capabilities onto the world. Nice to see Altman pointing out that he shares birthdays with Oppenheimer. I happen to wear a green fedora as did Oppenheimer; is that just a coincidence?*

When Sam Altman was fired as CEO of OpenAI in November 2023, one of the reasons circulating revolved around his lack of concern for safety.

> Recent changes in OpenAI's board should give us all more cause for concern about the company's commitment to safety. On the other hand, its competitor, Anthropic, is taking AI safety seriously by incorporating as a Public-Benefit Corporation (PBC) and Long-Term Benefit Trust. Artificial intelligence (AI) presents a real and present danger to society. Large language models (LLMs) like ChatGPT can exacerbate global inequities, be weaponized for large-scale cyberattacks, and evolve in ways that no one can predict or control. When Sam Altman was ousted from OpenAI in November, the organization hinted that it was related to his neglect of AI safety. However, these questions were largely quieted when Altman was rehired, and he and other executives carefully managed the messaging to keep the company's reputation intact.[118]

Although the interim CEO of OpenAI, Emmett Shear,[119] dismissed concerns that the board fired Altman due to safety concerns,

> the debacle should give pause to those concerned about the potential harms of AI. Not only did Altman's rehiring reveal the soft power he holds over the company, but the profile of the new board members appears to be more singularly focused on profits than their predecessors. The changes may reassure customers and investors of OpenAI's ability to profitably scale ChatGPT, but it should raise doubts about OpenAI's commitment to its purpose, which is to "ensure that artificial general intelligence benefits all of humanity."[120]

It is important to note that one of the other major players in building a god—Anthropic—became a start-up from employees who left OpenAI due to safety concerns:

---

* Yes, entirely coincidental. The green fedora was a Christmas gift from my in-laws. It is an antique fedora from Biltmore Hats—the factory where my father worked as the chief blocker for forty-two years. The very hat I was given quite possibly passed through his hands. If so, thanks, Pops.

Brother and sister Dario and Daniela Amodei left their executive positions at OpenAI to launch Anthropic in 2021. Dario had been leading the team that developed OpenAI's GPT-2 and GPT-3 models. When asked in 2023 why he left OpenAI, he could credibly point to the lack of attention OpenAI paid to safety, responsibility, and "controllability" in the development of OpenAI's chatbots.[121]

So there is definitely no lack of drama at the highest levels of the AI business. But aside from the soap opera escapades that were the OpenAI Board of Directors and the firing and rehiring of Sam Altman, it is important not to forget what is at stake here. And that is the degree to which AI existential risk is being taken seriously—or not. In the latter half of 2023, Altman managed to stage a world tour warning politicians in various countries to take action in developing policies, laws, and governance over emerging AI technologies. And yet, he's moving as quickly as possible to develop AGI. For the record, OpenAI has invested some of its resources into safety programs such as establishing safety systems, preparedness, and superalignment teams.[122] Meanwhile, over at Anthropic:

> The Amodeis were quite serious about baking ethics and safety into their business after seeing the warning signs at OpenAI. They named their company Anthropic to signal that humans (anthro) are at the center of the AI story and should guide its progress. More than that, they listed Anthropic as a public-benefit corporation (PBC) in Delaware. They join a rather small group of about 4000 companies—including Patagonia, Ben & Jerry's, and Kickstarter—that are committed to their stakeholders and shareholders, but also to the public good.[123]

Anthropic deliberately kept one of their large language models (LLMs)—Claude—back from public release in order to ensure that the information it provided could not be hacked or "red teamed"* to produce harmful advice. And although they are making efforts toward controllable and safe AGI, some are skeptical that they are on the right path:

> Some academic experts are concerned. . . . David Krueger, a computer science professor at the University of Cambridge and lead organizer of the recent open letter warning about existential risk from AI . . . thought Anthropic had too much faith that it can learn about safety by testing advanced models. "It's pretty hard to get really solid empirical evidence here, because you might just have a system that is deceptive or that has

---

* A "red team" involves a group of people who deliberately try to find and exploit weaknesses within an AI system. This helps companies identify and rectify these issues before they go to market.

failures that are pretty hard to elicit through any sort of testing. . . . The whole prospect of going forward with developing more powerful models, with the assumption that we're going to find a way to make them safe, is something I basically disagree with. . . . Right now we're trapped in a situation where people feel the need to race against other developers. I think they should stop doing that. Anthropic, DeepMind, OpenAI, Microsoft, Google need to get together and say, 'We're going to stop.'"[124]

The idea of slowing or simply stopping the building of an SMG in mid-form has been considered throughout AI communities for years. Some, like Eliezer Yudkowsky, believe we should forget about the idea that we will be able to control such a god and insist that we should shut the whole thing down:

> Many researchers steeped in these issues, including myself, expect that the most likely result of building a superhumanly smart AI, under anything remotely like the current circumstances, is that literally everyone on Earth will die. Not as in "maybe possibly some remote chance," but as in "that is the obvious thing that would happen."[125]

As one of the most experienced in the AI Risk business, Yudkowsky has been working on the alignment problem since 2001. And he's quite concerned that so few of us are doing anything about it:

> There's no proposed plan for how we could do any such thing and survive. OpenAI's openly declared intention is to make some future AI do our AI alignment homework. Just hearing that this is the plan ought to be enough to get any sensible person to panic. The other leading AI lab, DeepMind, has no plan at all.[126]

Others, such as Anthony Aguirre, executive director and board secretary at the Future of Life Institute, believe that we should control or limit the amount of power and capabilities that such gods will wield.

> AI engineers at some of the largest of our technology companies are racing to push these giant experiments in machine intelligence to the next level. We should not do so. Not now, perhaps not ever. Why? Because soon after these machine intelligence systems compete with human intelligence, we are likely to progressively lose control of them, and possibly even lose control to them. This . . . is an extended argument for why we should not, in the next few years, irrevocably open this gate: we should not train neural networks better than nearly everyone at a wide range of intellectual tasks, let alone ones better than the very best human experts or even all

of human civilization. Instead, we should set an indefinite hard limit on the total computation employed in training an individual complete neural network, a bound on how fast such a neural network runs, and likely other limits.[127]

In this way, Aguirre believes, we can "close the gate'" on such powerful machine gods so that they cannot get away from our control. And the best way to do so is to limit their computing power so that it becomes impossible for them to develop into highly intelligent rogue agents.

[A]t present we know of only one way to make such AI, which is via truly massive computations of deep neural networks. So all we have to do is to not do those incredibly difficult and expensive computations. However, since companies are currently racing each other to perform them under heavy competitive and financial pressure from investors and otherwise, it will require regulation from the outside to place this limit.[128]

Others, such as Justin Shovelain and Elliot Mckernon, believe that we might be able to "box" an AI system and limit its potential power (in theory if not in practice):

Can we box an AI while still being able to use it? We could try preventing it from physically interacting with its environment and only permitting it to present information to a user, but a superintelligence could abuse any communication channels to manipulate its users into granting it more power or improving its predictive abilities. To successfully box the AI in this manner, we'd need to constrain both its ability to physically interact with its environment and communicate information and manipulate. We'll call this output boxing: containing an AI by constraining its various outputs.[129]

It had always been my intention in trying to build the OSTOK Project to limit the amount of data that such a powerful form of AI could access. One has to wonder why OpenAI and other Big Tech companies are using the entire internet on which to train their large language models (LLMs). I just don't see the point of having computing power sift through information that is not essential for it to operate at peak performance. Does ChatGPT need to scour through every instance of pornography on the internet? If so, toward what end?

Boxing AI's capabilities, closing the gate on it, or shutting it down completely are just a few options to consider as we move into the future and AI capabilities continue to progress to ever more godlike powers.

We will consider these and others in chapter 5 when we look at what is currently taking place regarding the governance of AI. Getting politicians to agree on anything is difficult at best. It should come as little surprise, then, that this task has not become easier given its urgency to solve. But as we shall see in chapter 5, there is, actually, considerable movement in a number of countries regarding the governance and regulation of AI. But until we coordinate our efforts collectively and strategically, it still feels very much like an enormous undertaking—one that is far more layered and difficult than the Manhattan Project.

And so the race is on between the big fish in Big Tech—OpenAI, Anthropic, DeepMind, Meta, and Microsoft—to build an AI machine god first.* Others will join the race or be smaller fish helping one of the larger fish. But the race is on. And nobody knows whether we should stop it, redirect it, or make it go faster. The doomsayers or safetyists preach caution and prudence while the naysayers or effective accelerationists (or e/accs) say: "Faster, faster!" The doomsayers say AGI is inevitable and may arrive much sooner than anticipated with dire consequences. The naysayers say AGI will be entirely controllable and will only produce the very best for a company, a country, and the world.

Irrespective of what is going to happen, the need for critical thinking and ethical reasoning in a rapidly automating world is now. It has become a global moral imperative that we establish cooperative, transparent guidelines that can direct the safe and effective uses of AI technologies in all its forms and applications. It is important to start the conversation about AI at all levels of engagement and activity and to keep it going until we reduce the threat of an SMG to zero. Researchers, academics, technicians, computer programmers, engineers, politicians, and the general public all must be informed and engaged in the dialogue about the future of AI.

As we finish this first half of the book, it has been my hope to raise awareness and provide readers with enough reliable background information to confidently engage in dialogue about various aspects of AI. Now that we have more than enough information about the current developments of AI, we can move on to the second part of the book and consider some of the ethical implications, governance, and future of AI as humanity continues on its path to build an AI machine god.

---

* We also have to believe that major defense operatives globally are also involved in this race to some degree.

AI AND WHAT WE CAN DO ABOUT IT

II

Critical Thinking and the Ethics of AI

4

How to Build a God in Our Own Image

Whe have covered a fair bit of material regarding the history, benefits, and harms of AI. Now we need to think critically and ethically about what we should be doing about it. But what does that mean? How does one think "critically" and "ethically" about AI? And how will doing so help us better work through the difficulties we are, and will be, facing on the road to building a superintelligent machine god (SMG)?

The Importance of Critical Thinking

Critical thinking involves careful reflection on how and why you believe what you do. It consists of a skill set that allows us to better analyze and consider the value of our beliefs and those of others as they are measured by universally accepted criteria. Critical thinking is important because what you think often influences how you behave, which affects others whose beliefs and actions affect you. It increases our ability to think more clearly about important issues and empowers us with the capacity to be confident in what we believe. Critical thinking teaches us that there really are better and worse ways to think about important issues. In doing so, it focuses our thoughts to allow us to communicate more effectively and have "adult conversations" about important issues, disagree with one another, and still get along as civilized human beings. But perhaps most important, we must realize that *fairness* is the very cornerstone or Golden Rule of critical thinking. If we all abide by the rules of critical thinking and apply the skills fairly, we all get more of what we want. And this is one of the most

	Countries Abiding by Accord	Countries Not Abiding by Accord
Countries Abiding by Accord	WIN: All countries reap the benefits of AI	LOSE: All countries face Existential Risk
Countries Not Abiding by Accord	LOSE: All countries face Existential Risk	LOSE: All countries face Existential Risk

Figure 4.1. Prisoner's dilemma for global AI cooperation. *Created for the author. All rights to the author.*

important facts to remember as we move into an uncertain future where we must regulate and guide the path we want AI to take (figure 4.1).

For if all of the countries of the world abide fairly by the rules and regulations we eventually establish (and consider in chapter 5), then we all stand the best possible chance for benefiting from the positive advancements of AI technologies. But if a country or two decides to cheat rather than cooperate, well, that's not playing fairly, and that will lead perhaps all of us, including the cheaters themselves, to suffer greatly. And none of us as individuals or countries wants that; to do so would be illogical, irrational, and entirely counter to anyone and everyone's best wishes. In other words, to put it plainly, only a complete and total asshole would play unfairly; for it would mean harming themselves as much as the rest of the world. And we know that there's more than just a few of those types ruling over countries throughout the world at the moment. So we better get this right on the first try.

What, then, are the fair use tools in the critical thinking skill set that will make us all better problem solvers, decision makers, and responsible thinkers? For millennia, philosophers, mathematicians, logicians, scientists, writers, and many others have developed the critical thinking tools that allow all of us to improve our abilities to interpret, understand, critique, and act on information. I have taken a number of the most important tools and distilled them into an easy-to-apply skill set. The skill set of critical thinking allows us to better separate facts from feelings and acknowledges that there is value to our beliefs, our ideas, and our opinions. For our purposes, the ABCs of critical thinking that I mention in all of my books and lectures should give us a good account of their importance.

The first tool involves the construction of our ideas into formal *arguments* so that we understand our ideas better and can communicate them much more efficiently. So, *A is for Argument*. But what is an argument? An argument is always made up of two parts: your main point (what you believe to be true) and the reasons for believing it. In critical thinking, we call your main point the conclusion, and your reasons for believing it are called premises. In order to have an argument, you must have at least one premise supporting a conclusion. For example, if you believe climate change is really happening (your conclusion), what are your reasons (your premises) for believing this? If you think a particular movie is great (your conclusion), why? What are your premises (or reasons) for believing this? Once you begin providing reasons for why you believe what you do, you develop actual arguments. And this will help you think through your ideas and express yourself more clearly when it comes to understanding, interpreting, and acting on information about artificial intelligence. But how can we distinguish a good argument from a bad one? To help you to visualize this, think of an argument as a house (figure 4.2).

Its roof is the conclusion; its walls are the premises that support the roof; and the walls rest on a foundation made up of five universal criteria: consistency, simplicity, relevance, reliability, and sufficiency.

Whatever your conclusion is, your premises must be strong enough to support it as walls would with a roof. And the strength of the walls or premises is dependent upon their ability to satisfy the five universal foundational criteria.

Figure 4.2.   Argument as house. *Created by the author.*

And so, all arguments rest on a foundation of universal criteria that measure the value of an argument. If they completely satisfy all criteria, there is greater value to an argument than if the argument fails to satisfy some or all of them.

- **Consistency:** This is often referred to as the mother of all criteria when it comes to logic and critical thinking. All of our experiences have a uniformity or consistency to them. You don't put bread in a toaster and have it come up unicorns. Or internal organs. Or twenty-dollar bills. It either comes up toast, or it doesn't. There are no other options. This is consistent with our experiences. Once an individual violates laws of consistency, they have abandoned the possibility of producing a sound argument. So whenever you spot an obvious contradiction or inconsistency either through what somebody says or what they have done, the reason why it bothers you is because we are all hardwired with built-in consistency meters that we depend upon in order for the world to make sense. That's why, when we observe inconsistencies or contradictions, we take note of them. This skill—this capacity to notice such inconsistent behavior—is what kept our ancestors alive for so long. Observation of consistency of behavior allows all species on this planet to make predictions to promote the likelihood of survival and reproduction. Nature does not abhor a vacuum; it abhors inconsistency. So when someone's arguments or beliefs are inconsistent, we cannot be convinced by them.
- **Simplicity:** Sometimes referred to as parsimony, simplicity is universally valued as a criterion to support premises, because if we can present an argument that accomplishes just as much but with fewer premises, it tends to be more highly valued. Often referred to as Occam's razor—named after William of Ockham, a fourteenth-century Franciscan friar and scholastic philosopher—the criterion is often best summed up in his famous saying: "Do not multiply entities beyond necessity." What Ockham meant by this is that you really don't need more premises than are necessary to responsibly support your conclusion and satisfy the foundational criteria. One might even say that Ockham was the originator of the KISS Principle (Keep It Simple Stupid). In other words, when faced with competing arguments, the one that can say more with the fewest premises is generally favored.[1] So, for example, if at some point in the future, someone figures out a fairly simple solution

to controlling or containing the potential harmful effects of a superintelligent machine god, then so be it. Why would we need anything more complicated if a simpler solution works perfectly well?

- **Relevance:** It seems blatantly obvious that premises should be relevant to the support of a conclusion. And yet there are many ways in which people introduce irrelevant premises into an argument in the guise of presenting relevant points. The fallacy called a "red herring" is a fallacy of improper relevance. One commits a red herring fallacy when they attempt to circumvent addressing or responding to the topic at hand by discussing a subject or subjects that have nothing to do with the currently discussed topic.[2] The *ad hominem* fallacy also suffers from irrelevance. Calling out a person's personal attributes rather than focusing on their argument is not relevant to whether or not one should accept their argument.

- **Reliability:** The manner in which information is attained and used as premises to support a conclusion is extremely important. Attaining dependable, factual information is critical in the continued support of one's conclusion. Reliability is a universal criterion that demands that we attain our information from trustworthy sources, and that we do not shirk our responsibility in this regard. Many people around the world formulate opinions about extremely important issues based on information they find on Facebook or Twitter. We now know that such social media platforms are filled with false information from various sources in an attempt to sway public opinion.[3] As we move forward with various online manipulations such as deepfakes, we will need to be ever more vigilant in determining the factual truth of the information we come across. A greater need and emphasis, therefore, will be placed on ways in which we can assure the veracity or truthfulness of information.

- **Sufficiency:** We all have some idea of what it means for something to be sufficient. Sufficiency requires enough of something to satisfy some condition, want, or goal. For example, if I am in a busy metropolitan city like New York, and I need to get from one place to another, a taxi would be sufficient. I do not need an armored motorcade with limousine service (as nice as that might be) to get me down Forty-Second Street. In critical thinking, we often talk of premises providing the conclusion with enough evidence for support. But how much is enough? When are the premises sufficient in providing enough evidence to convince someone of your conclusion? The astronomer Carl Sagan once said, "Extraordinary claims require extraordinary evidence."[4] This basically means that the larger or more substantial your conclusion is, the more sufficiently convincing the evidence will need to be to support it.[5]

The degrees to which an argument satisfies or fails to satisfy these five criteria determine its value in critical thinking. If an argument is constructed well, then like a well-built house, it will rest firmly on the five foundational criteria mentioned earlier. And if it is not well constructed, its walls—or premises—will weaken, and the conclusion—or roof—will no longer be supported (figure 4.3).

You now know more about arguments and how they function in communication than the majority of the planet's population. Just remember, in the future, once you come to believe something, ask yourself what reasons or premises support that belief. And then consider to what extent those premises are consistent, relevant, unnecessarily complicated, reliably attained, and sufficient in supporting your belief.

The second tool in the critical thinking skill set is the acknowledgment of human biases. As we noted in the previous chapter, no matter who you are, or where you were born, or how you were raised, you cannot escape the numerous influences and constraints that have biased your thoughts and your actions throughout your entire life. So *B is for Bias*. A bias is a way in which people are influenced or constrained in the various ways they understand and act upon various types of information.

Figure 4.3.   Unsupported argument. *Created by the author.*

The manner in which we eventually come to acquire, revise, and retain opinions, beliefs, and attitudes about issues is the result of a long process of development, influenced by internal and external biases. Even before we were born, there were factors that would influence and bias the way in which we see and understand the world. We now need to become familiar with what these factors are not only to better understand how and why we—as humans—now believe the things we do, but also how and why AI technologies may be biased in ways that might produce negative outcomes.

There are many different types of biases. But they all generally fall under two categories: biological and cultural. Biological biases include factors such as genetic and epigenetic influences, neuropsychological factors, emotions, age, and health. Cultural biases involve constraints and influences such as family upbringing, ethnicity, religion, geographic location, friends, education, and the media. Everything you think and everything you do is the result of your biological and cultural biases. In learning how to think critically, readers are encouraged to be aware of their own biases and conduct a bias check to become aware that any new information presented to them must pass through a series of biased filters before they can accept it, reject it, or remain neutral to it. We often favor information that confirms our own biases—appropriately named confirmation bias. This is normal. And this is what it means in many respects to be human—that is, in seeking out information that makes us feel knowledgeable, confident, and secure, we need continued validation that what we believe is relevant, truthful, and worthy of being held in belief as a guide to our behavior. In critical thinking, however, we develop the ability to acknowledge and identify what our biases are in an effort to more fairly understand why it is we believe what we do and why it is that we act according to those particular beliefs. In this way, we become more critically reflective of our beliefs and can become fairer in our treatment of others whose opinions, ideas, and beliefs may differ from our own.

The same understanding of our own biases must extend to the development and control of all AI technologies. On one level, there will be those human agents who will choose to use such technologies for harmful and unethical ends: those who are motivated by their biases to make money, or to cause pain, ruin, and destruction to others, or perhaps simply because they wanted to see if they could—to see "what would happen next." Whatever the reasons for generating harm, we must become aware of the fact that some humans, some states, and some countries are going to use these technologies in ways their biases justify. And we need to be completely aware of this eventuality and be ready for when it comes in an effort to limit its impact.

The second type of harm from biases in AI technologies can come from conscious or unconscious programming of information into these advanced intelligent forms of AI. We noted earlier, in chapter 3, how some harms may arise from large language models (LLMs) as they develop their own biases due to the manner in which information is retrieved, categorized, and distributed. This can unfairly represent various marginalized groups.

The third type of harm from biases can simply come from the way in which specific algorithms behave. As we saw in chapter 3, there may be no intentional malice or harm initially in mind; but algorithms that continuously feed specific information to people who show interest in specific topics cater to the user's very biases in an effort to get more data and more money from that user. Every time you surf around online, information about you is being gathered so that more and more similar information can be sent your way to confirm your existing biases and please the emotional limbic system of your brain with a warm bath of neurotransmitters to reward you in your successful searches. So it is extremely important in critical thinking to understand the roles biases play in our abilities to interpret, understand, and act on information.

The final tool of the critical thinking skill set is context. When delving into the critical thinking aspect of context, we are essentially referring to elements such as time, place, and circumstance. Context serves as a lens through which we can gain a deeper understanding of the framework and background accompanying information. So, *C is for Context*. For a critical thinker, comprehending context provides a more extensive view of how information is situated and its ties to emotions, vested interests, and incentives. Within the realm of context, the significance of *fact-checking* becomes evident. If we neglect to accurately identify the pertinent facts related to information, or if we overlook the background of the information, or if we fall short in uncovering the who, what, why, where, when, and how behind it, our opinions on that information may be misguided, and our ability to assess it fairly diminishes significantly. It is crucial to recognize that within any culture, language is intricately woven into a context where individuals not only express their thoughts but also convey their biases. The setting in which these interactions occur often serves as the backdrop shaping the underlying information. A comprehensive understanding of context places us in a better position to grasp the reasons behind people's distinct thoughts and actions, which may diverge from our own. Numerous examples underscore the significance of context. An illustrative

Figure 4.4.   Mr. Spock. *Wikimedia Commons.*

instance comes from a vintage episode of *Star Trek* in the 1960s, featuring science officer Mr. Spock (figure 4.4) attempting to divert the USS *Enterprise* from its original course to his home planet, Vulcan. Initially, the crew is puzzled by Spock's seemingly erratic behavior. It is only revealed later that Spock must return to Vulcan within eight days to partake in the mating ritual known as Pon Farr, without which he would face his own death.

Much like a salmon swimming upstream for reproduction, Spock, with his pointed ears and the impassioned urgency of Pon Farr closing in, embarks on a journey across the galaxy in search of a mate. Understanding this context allows viewers to empathize with the motive behind Spock's unconventional actions. Without such context, passing a harsh and unfair judgment on Spock might be inevitable.

Although flattered when compared to a computer, Mr. Spock's behavior was understandable given the context in which it was situated. But what about the actual behavior of computers? Or superintelligent computers? Will we always be aware of the context in which they function and "make decisions" and "act"? We now know that this has created a problem in AI known as the "black box" or "interpretability problem." This problem refers to the challenge of understanding and explaining the decisions made by complex machine-learning models. As we noted in the first chapter, any advanced AI algorithms, particularly those using deep neural networks, operate as intricate and sophisticated systems with numerous layers of computation. Despite their effectiveness in making predictions or classifications, these models often lack transparency, which makes it extremely difficult for humans to understand how they arrived at specific outcomes.

The term "black box" refers to the fact that the internal workings of these AI models are hidden or opaque from us outside observers. This lack of transparency raises ethical concerns, especially, as we shall soon see, in critical applications such as health care, finance, or autonomous vehicles, where the ability to interpret and trust the decisions made by AI systems is crucial. Interpretability is essential for various reasons, including ensuring accountability, identifying potential biases in the model, and gaining user trust. Today, researchers and developers in AI are constantly working on establishing methods and techniques to increase the interpretability of machine-learning models, making their decision-making processes more transparent and understandable to human users. This is particularly important as AI systems become more prevalent in real-world applications, and the need for trust and accountability grows. Without understanding the context in which such superintelligent machines function, we will have a less likely chance of knowing how and why some systems behaved in manners not attuned to our commands or values. And such misfiring could potentially lead to some devastating effects.

Thinking Critically about Ethics

Now that we know how to formulate our ideas into arguments; and we are aware of the importance of understanding biases—ours, those of others, and those that emerge in AI; and we are also cognizant of the importance of understanding context, we are ready to use the skill set of our ABCs of critical thinking to generate arguments regarding the ethics of AI. But how do we think critically about the ethics of AI? In order to do so, we need to first understand that there are two types or categories of ethics regarding artificial intelligence. Allow me to explain.

Types of Ethics for AI

As we continue to develop various forms of AI technologies, there emerges the need to consider three very important ethical steps that need to be taken in guiding our actions so that we can attain the very best AI has to offer while, at the same time, limiting the significant harms it may generate. Generally, we need to think first about what types of ethical rules or precepts should guide human behavior in the use of such AI technologies as we move forward. Second, we need to think very carefully about what types of ethical commands we intend to give to such superintelligent forms of AI, which would allow us to control or contain them. And finally, we cannot forget that, should an AGI agent develop some level of sentience or consciousness, we have an obligation to consider the degree to which it must be afforded moral and legal rights. So the structure of our ethical reasoning would look something like this:

1. For Humans:

    a. What process will determine what ethical rules should guide the future of all human development and application of AI technologies up to and including those that cause existential risk? Within such a process, what doctrine of ethical precepts do we establish—perhaps one similar to a UN Charter of Human Rights—for all global parties to adhere to or abide by? And how should we determine what counts in deciding on ethical rules? What doesn't count? Who decides? And why?

2. For Artificial General Intelligence (AGI):

a. What ethical precepts do we establish that would be able to control the power of such a superintelligent machine god (SMG)? How should we determine these? What counts? And what doesn't? Who decides? Would something as simplistic as Asimov's Three Laws of Robotics* prevent an SMG catastrophe? Probably not. So then, we need to consider very carefully whether or not such precepts could control or contain an SMG; and if so, we then need to work out the very difficult tasks of determining how to do this. If we determine that no set of ethical guidelines can act as guardrails that can control an emerging SMG, then we have to move on to other options for control and containment such as technical or critical shutdown protocols.

b. What happens should an SMG become self-aware or conscious of its own existence? What if it develops agency and wishes to be autonomous? What if it could suffer? If any of these were to happen, would such an entity not automatically qualify for moral and legal rights?

As we saw earlier in the chapter, in order to think critically about AI, we need to establish sound arguments, acknowledge existing biases, and understand and appreciate contexts. But in order to reason ethically about AI, we need to understand some basic ethical theories, precepts, and principles that ground our arguments and provide structure and guidelines to our reasoning. In order to do so, we need to start at the beginning.

What Is Ethics?

Ethics is the study of how and why we value human behavior or actions. Values can be defined as good or bad, right or wrong, fair or unfair, just

---

* To restate Asimov's Three Laws of Robotics: "(1) a robot may not injure a human being or, through inaction, allow a human being to come to harm; (2) a robot must obey the orders given it by human beings except where such orders would conflict with the First Law; (3) a robot must protect its own existence as long as such protection does not conflict with the First or Second Law" (*Encyclopaedia Britannica Online*, s.v. "Three Laws of "Robotics," last modified April 25, 2024, https://www.britannica.com/topic/Three-Laws-of-Robotics).

or unjust. Morality, on the other hand, deals with the value of human actions in various settings (e.g., the mores or cultural rules practiced within cultures throughout the world). The field of ethics examines the nature of value as it may apply throughout all cultures. Context is extremely important in understanding the value of human behavior or actions because the setting and circumstances surrounding a particular set of actions often provide a more complete and comprehensive understanding of the motivation for such activity.

In one of the greatest philosophical works in history, Socrates refers to morality and ethics in the following way:

We are discussing no small matter, but how we ought to live.[6]

As we know from critical thinking, any argument requires a conclusion and a number of premises that support the conclusion. And we saw how the premises are the reasons one gives for maintaining a particular position. For example, consider the following question: Should Big Tech companies be allowed to race blindly and with full speed ahead in an effort to be the first to create an SMG when we currently have no proven measures in place to guide, control, or contain it? You may respond either in the affirmative or negative, but whichever side you take, you will need to support your conclusion with premises. If the reasons or premises for your conclusion are consistent, relevant, sufficient, and based on reliable information, then we may consider such a position to be well argued. But for an argument to be ethically justified, it needs something else: theoretical grounding. In other words, there are a number of different ethical theories, principles, and precepts out there. Which of them should be used when trying to modify both the behavior of humans and the behavior of a superintelligent machine god?

The Minimum Conception of Morality

Is there a basic or minimum conception of morality or moral action? Philosopher James Rachels defined it this way: "Morality is, at the very least, the effort to guide one's conduct by reason—that is, to do what there are the best reasons for doing—while giving equal weight to the interests of each individual who will be affected by one's conduct."[7] This gives us some indication of what it means to be a conscientious moral agent.

But I believe Rachels begs the question or is guilty of circular reasoning by assuming that we can make determinations as to the best reasons for guiding our conduct. Maybe we can accomplish this, but now, how should

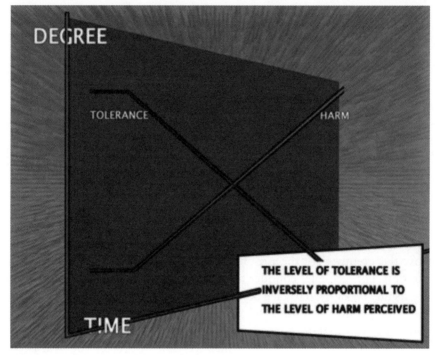

Figure 4.5.   The T-HIP model. *Created by the author.*

we determine the standards by which to measure their effectiveness? I believe there is a much more basic and simpler way of understanding the minimum framework from within which we make ethical judgments. It's called the tolerance-harm inverse proportion (or T-HIP) model (figure 4.5).

According to this model, human actions are tolerated as long as they produce very little harm. However, once one's perception of harm begins to increase, it reaches a common threshold as one's level of tolerance simultaneously decreases. At a given point in time, if one's perception of harm and one's tolerance intersect and inversely converge, we would consider the opposing of such behavior as justified. I have mentioned this model in other works.[8] Of course, the determination of what counts as "harm" is a separate issue. But once we have established an agreement that some level of harm is present, we can more readily see the relevance of and justification for our taking action due to our decrease in tolerance of the supposed harm.

The question you must now consider personally, dear reader, is how much advancement in AI technology should you or I, or we, the people, accept before its potential for harm decreases our tolerance to the point of acting? As we shall find in the final chapter, there are actions that can be taken to empower you in your abilities to do something about AI now, while it is being developed. Waiting to see what happens next is not an option we should tolerate. Waiting for others to take up the cause simply will not work either. The public must be well informed about how AI is shaping their world in real time so that they are given the opportunity to act if they find certain behavior intolerable. We might also want to say that those in the fields of AI have a moral duty to inform the public so they can help with the decision-making process required to move forward with the proper safe and fair governance of AI. To do so, we need to consider some of the ethical implications of advancing AI technologies. And in order to do this, we need some background regarding the theoretical foundations of ethical reasoning. These ethical theories, principles, and precepts will provide a more solid foundation to our arguments as we attempt to rationally work out the best ways forward in managing AI risk and governance.

Ethical Theories

There are a number of ethical theories and principles we shall examine prior to discussing various issues in AI. This is essential if we are to equip ourselves with the necessary means by which to discuss these issues intelligently and thoroughly. I will continuously stress this point: in taking part in ethical discourse, you cannot simply say anything you like and expect it to have equal cogency with good arguments. This is not *The Jerry Springer Show!* In order to better understand the complexities of the various issues, we need to understand the formal ethical theories that can be applied in considering what we believe are good reasons—or premises—for action.

Objectivity versus Subjectivity in Ethics

There's something that needs to be cleared up quickly when discussing ethics. There are those who believe that ethics is completely subjective and entirely up to an individual or group to decide upon the rules that govern human (and artificial life) behavior. And there are others who believe that ethical judgments can be objective. That is to say, there are some actions that can be objectively determined as being either good or bad, right or wrong. So who's right? Perhaps we can gain a better understanding of this particular aspect of the field of ethics by comparing it to scientific inquiry.

Scientific inquiry examines cause-and-effect relationships in the natural world. As such, scientific fields of study *describe* what occurs in the natural world. In this sense, science is *descriptive*—that is, it describes cause-and-effect relationships in the natural world. Ethics, on the other hand, *prescribes* how one ought to behave by making value judgments on specific forms of human behavior. Science describes cause-and-effect relationships in the natural world and presents us with *facts*. Ethics is *prescriptive* and prescribes how we ought to act and presents us with *values*. Science tells us what *is* the case regarding how humans behave. Ethics tells us what *ought* to be the case regarding how humans should behave.

In the various natural sciences, a certain level of objectivity can be reached that is unparalleled in any other fields of study. For example, experimental results can continuously be found and repeated under the same conditions by scientists in different parts of the world. Achieving the same results is what gives scientific fields of inquiry their clout and their value. We depend on science to produce objective results. When NASA sends people to the International Space Station, they want to make sure they arrive and return home safely. This could not be done if science were not objective and repeatable in very dependable ways.

In ethics, however, we just don't find the same type of rigor and objectivity that is found within the sciences. But that's not to hold ethics to be blameworthy. The philosophical field of ethics appears to be subjective by its very nature. And that is no fault of the field of ethics itself. Perhaps someday, ethics will become more objective—and if so, it will most assuredly get there through the use of the various sciences. But currently, we live in a world where it is us humans who must do the heavy lifting by trying to figure out what the best rules of conduct are that should guide us in our attempt to anticipate and control the emergent agency of a superintelligent machine god (SMG). So what are some rules of conduct? What follows are some of the most common and popular ethical theories, precepts, and principles for us to consider when critically thinking about ethics.

Cultural Relativism
When we look at the vast differences between so many world cultures, which all have different moral codes, we must ask ourselves to what extent ethics is merely the random walk of culture, where no culture's morality can be considered justifiably "better" than any other. If I visit a far-off country vastly different from my own that engages in actions I would not consider ethically acceptable, who's right? Who's wrong? Or should we even ask these questions? For example, in some countries and cultures, human rights are commonly violated—like in Russia, China, Iran, Saudi

Arabia, and, especially, North Korea. Should we simply accept the fact that all of ethics is reducible to one's geopolitical situation in the world? Can all of ethics simply be reduced to a form of "When in Rome, do as the Romans do"? Is ethics simply a matter of what's right or wrong or good or bad within a particular country? If the continuous violation of human rights in these and many other countries is perfectly acceptable within a given culture, then where's the harm? The harm comes from the suffering of humans. If you believe ethics is simply a matter of what others do within their own cultures, then how do you feel about slavery or human trafficking? Female genital mutilation? Or child labor? If you believe any or all of these actions are ethically wrong, then you're probably not a cultural relativist. Like the majority of people on this planet, such actions, such acts that violate the rights of humans, instill within you a lowering of your tolerance because the level of harm has exceeded your acceptable limit; and so you would feel justified in acting out against such actions.

Okay, so now the tricky part. Are there ethical theories, precepts, or principles that cut across all cultures and are not simply a product of a given culture's relative time, place, and circumstance? If so, what has been proposed? Let's take a look at some of the more common ones.

Religion and Morality

Since religious beliefs have proposed ethical systems for millennia, could there be concepts and ideas embedded within some of them that may help us determine our best course of ethical action? Many religions believe in a divine command theory, which involves an all-loving creator who cares most for his greatest creation: humankind. And many believe further that such a being has bestowed upon us, his greatest creation, a list of dos and don'ts, which, for better or worse, will determine our favor with such a being. If we "choose" good over evil, the deity in question rewards us. And if we "decide" to ignore or violate such supernatural governing rules, well, then things don't look as good for us.

But of all the world's religions, which are the best rules, and how would we decide? For example, in Christianity, some of the key ethical principles include loving thy neighbor, forgiveness, humility, compassion, and so on. Like so many other religions, it also maintains the Golden Rule: "Do unto others as you would have them do unto you." In Islamic faiths, we find the Five Pillars of Islam: *Shahada* (faith), *Salat* (prayer), *Zakat* (obligatory charity), *Sawm* (fasting), and *Hajj* (pilgrimage), along with appeals to justice, compassion, and honesty. Islam, too, preaches a formulation of the Golden Rule: "None of you truly believes until he wishes for his brother what he wishes for himself." Within Judaism, some of the key ethical principles include the Ten Commandments, as well as the concepts of justice (*tzedek*), compassion (*chesed*), and righteousness. The Judaic formulation of the Golden Rule is the following: "Love your neighbor as yourself." In Hinduism, you will find key ethical principles of *Dharma* (duty/righteousness), karma (action and its consequences), *ahimsa* (nonviolence), and compassion. Though various formulations of the Golden Rule appear throughout Hindu writings, this is a common iteration: "This is the sum of duty: do not do to others what would cause pain if done to you." In Buddhism, we find the Four Noble Truths (the truth of suffering, the truth of the cause of suffering, the truth of the end of suffering, and the truth of the path that leads to the end of suffering), Eightfold Path (right view, right intention, right speech, right action, right livelihood, right effort, right mindfulness, and right concentration), and principles for compassion and nonharming. Some have proposed that incorporating the values of Buddhism in AI ethics will work if we keep in mind the potential for reducing suffering:

> Buddhism proposes a way of thinking about ethics based on the assumption that all sentient beings want to avoid pain. Thus, the Buddha teaches that an action is good if it leads to freedom from suffering. The implication of this teaching for artificial intelligence is that any ethical use of AI must strive to decrease pain and suffering. In other words, for example, facial recognition technology should be used only if it can be shown to reduce suffering or promote well-being. Moreover, the goal should be to reduce suffering for everyone—not just those who directly interact with AI.[9]

Like the previously mentioned religions, a formulation of the Golden Rule is found in Buddhist teachings: "Hurt not others in ways that you yourself would find hurtful." We see similar precepts in Sikhism, where the key ethical principles involve *Naam Japna* (meditation on God's name), *Kirat Karni* (honest living), *Vand Chakna* (sharing with others), and the value of compassion. The Golden Rule formulation for Sikhs is to "treat others as thou wouldst be treated thyself."

In Jainism, some of the key ethical principles include *Ahimsa* (nonviolence), truthfulness, non-stealing, chastity, and non-attachment. The Jain formulation of the Golden Rule is this: "In happiness and suffering, in joy and grief, we should regard all creatures as we regard our own self." And finally, for Confucianism, some of the key ethical principles include *Ren* (benevolence), *Li* (ritual propriety), *Xiao* (filial piety), *Yi* (righteousness),

Figure 4.6. The Golden Rule. *https://www.scarboromissions.ca/wp-content/uploads/2023/07/GR-poster-scaled.jpg.*

and *Zhi* (wisdom). The Confucian formulation of the Golden Rule is this: "Do not impose on others what you do not wish for yourself."

There are many different iterations of the Golden Rule woven into the belief structures of many belief systems—both religious and secular (figure 4.6).

Although there are several commonalities of ethical principles between world religions—such as the Golden Rule—how would we be able to determine which of the central tenets or principles of each of them should be considered valuable in determining ethical behavior with AI? We may find similar qualities in less religiously based codes of ethics. Let's consider a more secular-based way of dealing with ethical behavior as a form of mutual consideration and respect.

Thomas Hobbes: The State of Nature and the Social Contract

Commonly referred to as the father of modern philosophy, Thomas Hobbes was born in Malmesbury, Wiltshire, England, on April 5, 1588. Apparently, Hobbes was born prematurely when his mother had learned of the impending approach of the Spanish Armada coming up the English Channel. "Fear and I were born twins" was one of Hobbes's most celebrated descriptions not only of his birth but of his views that fear of death and the need for security were strong influencing factors in the development of moral and political systems between each other and other states.

Hobbes has made significant contributions to philosophy, but it is his work *Leviathan* that has brought Hobbes his greatest recognition. For it is in this work that Hobbes mentions the state of nature and his concept of social contract theory. These two ideas are, in many ways, both accurate and functional in societies today and perhaps especially when dealing with the ethics of artificial intelligence. Hobbes looked carefully at society and compared the nation or state to an organism like a large person. Every part of the state parallels the function of the various parts of the human body. Since humans are the creators of the state, the first part of his project is to describe human nature. To Hobbes, this was obvious:

Human Nature = All human actions are performed for the benefit of the individual.

In other words, no matter how unselfish or charitable an act may seem, we always do it because we somehow gain from it. In terms of ethics—that is, how we should act toward one another—Hobbes believed that this must take into account the fact that we are incredibly selfish.

THE STATE OF NATURE

Consider the following thought experiment. Imagine a society where there is no government or ruling class but, instead, a state where it's everyone for themselves. It is not difficult to do this. In such a state, Hobbes imagined that, given the general equality of humans in strength and intelligence, there would have been situations that would have made our ancestors naturally prone to quarrel. Hobbes lists three natural causes for quarreling among people:

1. The competition for limited supplies and material (e.g., food, tools, clothing, etc.).
2. Distrust of one another.
3. The glory sought in the attempt of people to maintain their reputation of power and control.

Since humans are naturally inclined to quarrel in these ways, Hobbes saw the natural conditions of humans as a state of perpetual war of all against all—a place of insecurity without morality in which everyone lives in constant fear of death:

> In such condition, there is no place for industry, because the fruit thereof is uncertain; and consequently no culture of the earth, no navigation, nor use of the commodities that may be imported by sea; no commodious building, no instruments of moving and removing such things as require much force; no knowledge of the face of the earth, no account of time, no arts, no letters, no society; and which is worst of all, continual fear and danger of violent death; and the life of man, solitary, poor, nasty, brutish, and short.[10]

It is interesting to see how Hobbes, in some ways, anticipates the works of Charles Darwin by stating that the desires and passions that all humans have—and the actions that ensue—are not evil in and of themselves until they know a law that forbids them. Hobbes is referring here to the need to instill some type of order that will rid us of our fear of death and insecurity.

> To this war of every man against every man, this also is consequent; that nothing can be unjust. The notions of right and wrong, justice and injustice have there no place. Where there is no common power, there is no law; where no law, no injustice.[11]

Hobbes divides the human faculties into two types: the passions and reason. Although our passions are entirely self-serving, our fear of death and need for security can rationally incline us all toward peace. Hobbes sees this rational desire as being necessary to commodious living. Reason provides us with the ability to develop articles of peace such as agreements. These articles Hobbes calls the laws of nature.

THE LAWS OF NATURE

To secure peace, Hobbes uses the language of the natural law tradition of morality led by Dutch politician Hugo Grotius (1583–1645). To Grotius, moral precepts can be found in universal principles, which are fixed in nature itself. Hobbes believed one could derive such laws, referred to as "laws of nature," by recognizing our own selfish nature and desire for social agreement. To Hobbes:

> A Law of Nature (*Lex Naturalis*) is a precept, or a general rule, found out by reason, by which a man is forbidden to do, that, which is destructive of his life, or taketh away the means of preserving the same; and to omit, that, by which he thinketh it may be best preserved.[12]

Hobbes establishes fifteen natural laws, but those that are most noted and practiced are his first three. It is important to realize that Hobbes's ethical system is axiomatic—that is, due in no small part to his influence with mathematics. Hobbes has presented what he believed to be fundamental truths about human nature, in other words, that we are a greedy self-serving bunch who constantly live in fear of losing our possessions and life and constantly desire security. From these truths—these axioms—he wants to deduce a system of morality and politics that is in keeping with our human nature. And so he establishes his laws of nature.

His first law of nature plays on the human desire to find security or peace:

> That every man ought to endeavor peace, as far as he has hope of obtaining it; and when he cannot obtain it, that he may seek, and use, all helps, and advantages of war. The first branch of which rule containeth the first, and fundamental law of nature; which is, to seek peace and follow it. The second, the sum of the right of nature; which is, by all means we can, to defend our selves.[13]

Seeking peace or security, however, does not come without some type of sacrifice. If peace is truly in the best interest of all concerned, then we must forgo some of our liberties or rights in order to ensure that security is attained. Hobbes states that from this first law of nature, there is derived a second law:

> That a man be willing, when others are so too, as far-forth, as for peace, and defence of himself he shall think it necessary, to lay down this right to all things; and be contented with so much liberty against other men, as he would allow other men against himself. For as long as every man holdeth this right, of doing any thing he liketh; so long are all men in the condition of war. But if other men will not lay down their right, as well as he; then there is no reason for any one to divest himself of his: For that were to expose himself to prey (which no man is bound to) rather than to dispose himself to peace. This is that law of the Gospel; Whatsoever you require that others should do to you, that do ye to them.[14]

Notice how the Golden Rule has been baked right into Hobbes's second law of nature. This renouncement or transference of rights leads us to be obliged or bound not to hinder those to whom such a right is granted. And so we owe these people a duty not to interfere. But notice also that the motivation to transfer rights is ultimately selfish; in other words, neither of us wants a state of nature without peace or security, so we are both going to be willing to transfer some of our rights to ensure this. I agree not to kill you or steal your possessions, and you agree not to do the same to me. And so we enter into an agreement, or bond, or mutual transferring of rights known as a social contract.

Hobbes's third law of nature refers to the fact that we have an obligation to keep our agreements.

> From that law of Nature, by which we are obliged to transfer to another, such Rights, as being retained, hinder the peace of Mankind, there followeth a Third; which is this, *That men perform their Covenants made*: without which, Covenants are in vain, and but Empty words; and the Right of all men to all things remaining, we are still in the condition of War.[15]

But what value is an agreement or covenant if someone decides to cheat? Why should we bother to seek peace, fairly practice the Golden Rule, and keep our agreements if others are not willing to do so? This is something we will need to consider very carefully as we move forward in establishing AI governance policies for individuals, industry, and nations. To Hobbes, the answer can be found in the establishment of a system that honors contracts that are binding and consequential.

SOCIAL CONTRACT THEORY

Thomas Hobbes knew that people would always be tempted to break their contract of rights transference. And so he needed to include an appeals process in which one could seek justice should a contract with another be violated. Thus, he introduced the *political sovereign*. This is a person or collective body that has the power to punish those who violate their contracts with others. Again, the motivation for appointing a sovereign is for selfish reasons; in other words, none of us wants to lose our state of peace and security, but for fear of losing this, we will play it safe and appoint and pay someone with the power to punish those who violate their contracts of transference of rights. So we surrender our right to govern ourselves to a single person or assembly.

> [I]t is a real unity of them all, in one and the same person, made by covenant of every man with every man, in such manner, as if every man should say to every man I authorize and give up my right of governing my self, to this man, or to this assembly of men, on this condition, that thou give up thy right to him, and authorise in like manner . . . this is the generation of that great Leviathan, or rather (to speak more reverently) of that mortal god, to which we owe under the immortal god, our peace and defence . . . and he that carryeth this person, is called Sovereign, and said to have sovereign power, and every one besides his subject.[16]

Hobbes's descriptions of the state of nature, the laws of nature, and the social contract paint an interesting picture of us humans. Much of what he says is not only true, but his ideas have been incorporated today into many of our legal systems. Just think for a moment of what you would do if someone wronged or harmed you in some way. You could call the police and have the person arrested, taken to trial, sued, and perhaps sent to prison—and why? Because that person did not play fairly—that is, they broke the laws that have been established as contracts between us all. This is known as the rule of law—an agreement whereby no person, group, corporation, or country has greater entitlement above the law itself. Without laws, and the capacities by which to enforce them, we would be living in a Hobbesian state of nature not unlike the Wild West or law of the jungle where anything goes and we are all left to defend ourselves. Does anyone truly want to live in such anarchic dystopias? It would be irrational to favor such nightmarish scenarios as they are simply not conducive to healthy, productive living.

Constantly living in fear due to lack of security is a horrible way for humans to exist. But it is in exactly such a state that we may find ourselves if we don't get to work on a globally accepted accord. In order to avoid such dystopian states of being, we need to collectively get very Hobbesian very quickly. We are quickly facing a situation of either collectively working together to assure mutual benefits from AI for all countries of the world; or facing the very real possibility that some individual, company, or country will go rogue and decide to use AI technologies for their own benefit and to the potential suffering of others and other species. And at that point, the tolerance-harm threshold will have been breached. We must therefore prepare ourselves for potential scenarios in the future that will involve noncooperative parties, and we must think now about how such global infractions should be penalized.

Now that we have looked at a number of ways that might influence how we should establish ethical principles and precepts, let's consider an ethical theory that maintains that it's all about the consequences when deciding what's right and what's wrong.

Utilitarianism

This type of ethical theory belongs to a class of ethical theories called "consequentialism" because it deals with consequences that result from specific actions. In a phrase, we can consider it in terms of the ends justifying the means. The ethical theory of utilitarianism was first proposed by David Hume (1711–1776) but received its definitive formulation both by Jeremy Bentham (1748–1832) and John Stuart Mill (1806–1873). During these historic times, there was a great deal of political change occurring throughout Europe and the New World (America). The 1600s saw the separation of church and state, as well as the development of parliamentary proceedings rather than rule purely by the monarchy. From the Renaissance, to the Enlightenment, and through the Industrial Revolution, new ideas were being developed. It is not surprising to see this occur in ethical thinking as well. Jeremy Bentham had maintained that morality had little to do with pleasing any God or upholding abstract rules but with bringing about the greatest happiness for the greatest number of people. This is what should motivate people's actions.

Jeremy Bentham was a true polymath who was way ahead of his time in a number of ways.* He was an abolitionist to slavery, a champion of human rights, an advocate for women's rights, a prison reformist, an animal rights activist, and a liberator of human sexuality and the decriminalization

---

* It is by no accident that my son, Jeremy, received his name largely from the influence of this great man (and also because of a really great Pearl Jam song).

Figure 4.7. Jeremy Bentham. *Photo taken by author.*

of homosexuality; he helped develop the establishment of the University College London (UCL), and he was the first person in history to donate his body to medical science. Knowing that grave robbery was rampant throughout the United Kingdom, Bentham wanted to stop the illegal and unsavory occupation of digging up freshly buried corpses to be sold to university medical schools as cadavers. The practice became so popular that men had to be hired to guard over cemeteries through the late-night hours—an occupation that gave rise to the phrase "working the graveyard shift." Bentham requested that his body be used for whatever purposes the medical school professors saw fit. But he then asked, if not too much trouble, to have his skeleton put back in his clothes and be placed in a specially built cabinet so that he may preside over the halls of UCL and occasionally get wheeled out during various festivities (which has been honored to this day) (figure 4.7).

Oh yes, and he was the central founder and champion of the ethical theory of utilitarianism. So what does this ethical theory, known as the principle of utility, propose? Whenever any member of society has the choice between alternative actions or social policies, they should choose that which has the best overall consequences for everyone concerned. So the rightness or wrongness or goodness or badness of human behavior can be determined in relation to the extent that they either increase or decrease human well-being or "utility."

> Bentham defended an objective form of morality that could be measured in a scientific way. As an empiricist, he came up with a way to "weigh" or quantify pleasures and pains as the consequences of an action. He called this set of metaphorical scales the "hedonic" or "felicific calculus," allowing a rational moral agent to think through, and then act on, the right—moral—thing to do. The "hedonic calculus" is used to measure how much pain or pleasure an action will cause. It takes into consideration how near or far away the consequence will be, how intense it will be and how long it will last, if it will lead on to further pleasures or pains, and how certain we are that this consequence will result from the action under consideration. The moral decision maker is meant to act as an "impartial observer" or "disinterested bystander," to be as objective as they can be and choose the action that will produce the greatest amount of good.[17]

Contemporary philosopher Peter Singer is a utilitarian who argues not only for the rights of animals but also for the rights of AI—should it develop sentience. In a recent interview with Dr. Singer on my podcast *All Thinks Considered*,[18] he mentioned Bentham's main precept regarding the treatment of animals: "The question is not 'Can they reason?' nor 'Can they talk?'—but 'Can they suffer?'"[19] I asked Singer about AI and to what extent we will need to consider the rights of emerging forms of intelligence. He stated: "When robots become conscious; when they also, like humans and animals, become capable of suffering or of enjoying their lives, they certainly should have rights."[20] We shall return to some of Singer's utilitarian insights regarding the potential for the emergence of consciousness in AI systems.

For now, it's important to understand that the main idea driving the ethical theory of utilitarianism is that we should act in ways that generally benefit the greatest number of those affected by our actions. So, when pondering the ethics of AI and how it will affect us in all its various forms, we should consider the consequences of our actions and the degree to which they contribute to the overall benefit or suffering of those who will be affected—including humans and all other species as well. But are consequences all that matter in ethics? Some might disagree.

Deontology

The ethical theory of deontology involves the ethical duty and motivation to act according to rules or universal laws. The term "deontology" is derived from the Greek word *deont*, or "duty," and *logos*, or "word, thought, principle, speech, plan, or message." In other words, such a theory is not concerned so much about consequences because determining these may be rather subjective and arbitrary. What matters most is the will and motives that drive people to follow universal ethical laws like "Don't steal," "Tell the truth," and "Don't cheat."

Consider the following thought experiment: Imagine that you have the opportunity to steal a large amount of money from your employer. There's very little chance that you would get caught and you could really use the money for all types of expenses like paying down bills or giving to a charity. Should you take the money? According to German philosopher Immanuel Kant (1724–1804), you should never steal, even in cases such as these. To Kant, morality consists of following to the letter absolute rules that admit no exception. And so stealing, no matter what the surrounding circumstances, or what the consequences, is never morally justified. Kant did not appeal to the consequences of an action or any particular God's commandments for such reasoning. Instead, he appealed to reason itself.

Kant observed that, in life, there are many ways in which the word "ought" is used in a nonmoral context. If one wants to become a better tennis player, one ought to practice; if one wants to do well in school, one ought to study the material and attend classes. Kant referred to these as "hypothetical imperatives" because they tell us what to do provided that we have the relevant desires. If one does not wish to become a better tennis player or do well in school, one can simply choose not to practice or study the material. The only "binding force" in this imperative depends on one's desire. So one can easily ignore such obligations (and we do this all the time with weight loss, keeping promises, New Year's resolutions, etc.).

Unlike personal obligations, which are hypothetically imperative, Kant believed that moral obligations are *categorically imperative*. That means one *ought* to do X, period. Such moral rules do not cater to personal taste and preference. Instead they are universally binding. Where hypothetical "oughts" are possible because we have desires, categorical "oughts" are possible because we have reason and we have will.

> Kant held that nothing is good without qualification except a good will, and a good will is one that wills to act in accord with the moral law and out of respect for that law rather than out of natural inclinations. He saw the moral law as a categorical imperative—i.e., an unconditional command—and believed that its content could be established by human reason alone. Thus, the supreme categorical imperative is: "Act only on that maxim through which you can at the same time will that it should become a universal law." Kant considered that formulation of the categorical imperative to be equivalent to: "So act that you treat humanity in your own person and in the person of everyone else always at the same time as an end and never merely as means."[21]

That is to say, such universal laws are binding on us human agents because they are rational. Such categorical imperatives are rational because they are derived from a principle that every rational person must accept: that we should act only according to an ethical rule that we would want to become a universal law. Known as the first formulation of the categorical imperative, Kant believed that at any time we are about to decide on how to act, ask yourself whether or not the rule you are about to follow should be followed by everyone all the time who are faced with the exact same situation (this would make it a "universal law"). If so, then the rule is to be followed and the act permitted. So we would really want everyone not to steal, cheat, and lie. And because we are all in this together, we should all abide by the same categorical imperatives. In his second formulation of the categorical imperative, he states that we should never treat people as a means to an end but, instead, treat them with dignity and autonomy. Because we all have goals and projects, we should never use people along the way in order to attain them. People are not objects to be used and disposed of. They have intrinsic value, dignity, and worth. They are rational beings, and we must respect this rationality. In this way, it is quite different from the consequentialist results of utilitarianism.

In general, utilitarian and deontological ethical theories are often at odds with one another. To illustrate how this is so, consider the following thought experiment:

Imagine you are in a large city. Coming down a hill on a busy street is a runaway trolley (figure 4.8). It's heading right for five workers who are too busy working on the tracks ahead to notice the runaway train. It's too loud to warn them. But there is a switch that would redirect the trolley down another track. However, there is one worker who is on that track. If you pull the switch, one worker will die. If you do nothing, five workers will die. Would you pull the switch?

The results—that is to say, the consequences—seem pretty straightforward: You can do nothing and five people will die; or you can pull a switch causing one person to die. The so-called choice is yours. Now let's modify this thought experiment a bit. Suppose everything is the same with the runaway trolley heading for five men; but this time, you're on top of a bridge overlooking the tracks (figure 4.9). However, standing right next to you is a rather large gentleman. Since it is still too loud to warn the five workers, and you don't have enough time to run and pull the switch, you

Figure 4.8. Trolley example (1). *Created for the author. All rights to the author.*

Figure 4.9. Trolley example (2). *Created for the author. All rights to the author.*

could just push the large man onto the tracks and his body would stop the train and save the five workers. Would you push the large man to his death?

Unsurprisingly, the majority of respondents say that they would pull the switch but would not push the man. How about you? How and why did you come to your conclusion? Most people feel as though the switch somehow disconnects them from "causing" a death whereas the physical act of pushing someone to their death feels "too personal." But the results would be exactly the same in both cases: one man would be dead, and five would be saved. In the first scenario, most people are utilitarian (i.e., it's clearly better to save five lives than one). But in the second scenario, most people suddenly become deontologists and would instead consider the dignity and autonomy of the man next to them on the bridge and choose not to use him like an object or as a means to an end.* The purpose of such thought experiments is to make us think carefully about our behavior and reflect on what values we hold most dearly. It's not as though we will ever find ourselves in such situations; that is to miss the point. Such thought experiments put us in challenging and difficult mental places that deliberately make us feel uncomfortable. And that's good; because much of what is coming with new AI technological advancements is going to make us uncomfortable and, potentially, extinct. So we better get thinking about

---

* I should mention that when I asked noted utilitarian Peter Singer this question, he unhesitatingly said he would push the man to his death.

it sooner than later. As we will see in the next chapter on AI governance, much of the work we do at Convergence Analysis is based on "scenario research." But the scenarios we consider could actually occur; and so we need to consider their likelihood so that we can develop safe and thoughtful governance in a proactive rather than reactive manner. Remember, the overarching focus of our agency and every similar AI governance agency is to develop guidelines that will ensure that we receive the very best that AI has to offer the world, while limiting its potential for harm and suffering.

Virtue Ethics

Think for a moment about heroes, icons, and role models. When I was young, I had many heroes—each for various reasons: Spider-Man had the ultra-cool ability to swing from building to building in a relatively cool-looking suit; Batman did his best work in an eerie cave but with high-tech computers; Tarzan lived among the animals and could communicate with them—he could run, swing, and swim with the best of his primate brethren; Zorro was a champion of the exploited, had a cool outfit, and was good with a sword; and Superman was really just the best of them all—for, if not for a bit of Kryptonite, he was the most powerful by far.

Now, ask yourself this question:

Why was Superman good?

He certainly didn't have to be, but he "chose" to be and to do good for truth, justice, and the American way (even though he was conceived by a Canadian). But the bottom line is, he did not *have* to be good. As mentioned earlier, the Crash Test Dummies eloquently state in their "Superman Song": "Hey Bob, Supe had a straight job, even though he could have smashed through any bank in the United States. Well, he had the strength, but he would not."[22] So why didn't he? What virtues and principles kept Clark Kent on the straight and narrow? And what virtues and principles should guide either humans with very powerful forms of AI or the very powerful forms of AI themselves? When Stan Lee introduced us to Spider-Man in Marvel comics back in 1962, it was Peter Parker's Uncle Ben who told him: "With great power comes great responsibility."[23] This concept has been found throughout written history. We see it in the first century BCE parable of the Sword of Damocles and the medieval principle of "noblesse oblige," in other words, that power cannot simply be enjoyed for its privileges alone but necessarily makes its holders morally responsible both for what they choose to do with it and for what they fail to do with it.[24]

I believe heroes are important to any society because they reveal to us, through the actions of others, special qualities that we believe are good and wish to emulate as a collective group. It is perhaps for this reason that we are so surprised when virtuous people in positions of power, mentors, and heroes are caught acting unethically. We expected more of and from them. And they let us down.

Socrates, Plato, and Aristotle were among the first to consider ethics on the basis of quality of character. In the *Nicomachean Ethics* (325 BCE) Aristotle asked: "What is the good of man?" and answers by saying "an activity of the soul in conformity with virtue."[25] And so according to the ancient philosophers, to understand ethics is to understand what makes a person virtuous. Prior to the 1600s, a great deal of emphasis in the study of ethics went into trying to figure out what made good people, well, good. To the ancient philosophers, it was virtue. Throughout the Christian medieval period, it was to follow the divine law of God.

> After the Renaissance period (1400–1650), moral philosophy again became more secular, but philosophers did not return to the Greek way of thinking. Instead, the Divine Law was replaced by something called the "Moral Law." The Moral Law, which was said to spring from human reason rather than from God, was a system of rules specifying which actions are right. Our duty as moral persons, it was said, is to follow those rules. Thus, modern moral philosophers approached their subject by asking a question fundamentally different from the one asked by the ancients. Instead of asking *What traits of character make someone a good person?* they asked *What is the right thing to do?* This led them in a different direction.[26]

According to Rachels, this change in tack led philosophers to develop ethical theories based not in virtue but in rightness and obligation:

- **Ethical Egoism:** Each person ought to do whatever will best promote his or her own interests.
- **Social Contract Theory:** The right thing to do is to follow the rules that rational, self-interested people would agree to follow for their mutual benefit.
- **Utilitarianism:** One ought to do whatever will lead to the most happiness.

- **Kant's Theory:** Our duty is to follow rules that we could accept as universal laws—that is, rules that we would be willing for everyone to follow in all circumstances.[27]

So how did the ancient philosophers define virtue and develop an ethical theory based on it? Let's consider how virtue has been defined. Aristotle stated that a virtue is a character trait manifested in habitual action (i.e., that the trait is not fleeting but continues on a regular basis). Additionally, the virtue must differ from vices based on our preferences or avoidance (i.e., we admire certain virtuous characteristics of some people and loathe viceful characteristics in others). Thus "a virtue is a trait of character, manifested in habitual action, that it is good for a person to have."[28] In other words, in order for a person to be considered virtuous in some capacity, they cannot simply be virtuous on a single occasion but must repeatedly demonstrate this behavior over and over again. Hence, we come to rely on this characteristic trait within an individual.

And it is important to realize that there are many different virtues found in the human experience. They range from benevolence to fairness, to reasonableness, self-discipline, compassion, justice, loyalty, moderation, charity, courage, and so on.* In other words, there are a lot of human virtues. So how do we act on them in an ethical system? For this, Aristotle devised a tool he called the "golden mean" (figure 4.10): A virtue is "the mean by reference to two vices: the one of excess and the other of deficiency."[29] Imagine a linear scale where on the extreme left side stands the vice of deficiency and on the extreme right side stands the vice of excess. Right in the middle of these two extremes lies the golden mean of behavior. For example, consider what would define a person as being "courageous." According to Aristotle, it would be one who does not run away from danger but who also does not take unrealistic or foolhardy risks. In other words, they're somewhere in the middle or the mean between the two extremes.

We find virtues important in society because they tell us a great deal about an individual's character. And from that knowledge we are more secure in trusting individuals because they have repeatedly demonstrated their virtuous behavior. But on their own, virtues have pragmatic benefits as well. We value fairness because it distributes wealth and opportunity. Loyalty leads to stability and dependability among friends, family, and others. Open-mindedness and curiosity lead to tolerance, acceptance, and

---

* You can view a more complete list in appendix B.

COWARDICE   **COURAGE**   **OVERCONFIDENCE**

VICE OF <u>DEFECT</u>   VICE OF <u>EXCESS</u>

Figure 4.10.   The golden mean. *Created for the author. All rights to the author.*

humility. And these virtues continue to lead to longer and more trusting relationships within civil society.

It is important to note that no single ethical theory mentioned here or found anywhere in the world is sufficient to provide a complete and accurate guide for moral behavior. The importance of ethical theories lies in their capacity to provide foundational support for our arguments. And it is often the case that, in order to deal with moral and ethical issues in AI, we will need to choose the very best information from a variety of ethical theories. This is what the collective ethical program of principlism attempts to accomplish.

Principlism

This collective of ethical theories combines the principles of benefit (benevolence)/no harm (non-maleficence), autonomy (liberty, free will, and agency), and justice (social distribution of benefits and burdens). These principles were initially formalized as a moral decision-making approach by the National Commission for the Protection of Human Subjects of Biomedical and Behavioral Research in the Belmont Report on April 18, 1979.[30] The roots of the report trace its origins back to post–World War II military tribunal hearings against twenty-three German physicians for crimes against humanity. This led to the creation of the Nuremberg Code, which established standards for all future research subjects and advocated for informed consent for those participating in research on human subjects. The Belmont Report maintains that these principles have been established in order to be consistent with and not to conflict with many forms of ethical, theological, and social approaches toward moral decision-making.

When considering the ethics of artificial intelligence, we must construct sound arguments by anchoring them with relevant ethical precepts, principles, or theories. How we wish to do this is what awaits us. For example, in reference to having countries agree to play fairly according to universal laws of conduct, if you believe a formulation of the Golden Rule applies, such as "Do unto others as you would have them do unto you," this seems basic and simple enough. However, by itself, it is not enough to effectively control the behavior of humans, let alone the behavior of artificial life forms like powerful, superintelligent gods. It is a helpful guideline in general and may be used in conjunction with a series of other precepts, rules, or theories. The Golden Rule suggests that we should treat others as we want to be treated. However, it has some provisions. One key provision is that you can't cause harm to others and justify it by claiming you'd be okay with the same harm happening to you. For instance, you can't use AI to harm others and try to justify it because you'd be fine with being a victim of someone else's use of AI against you. This is where the Golden Rule falls a bit short, as it allows individuals like murderers, thieves, rapists, and generally harmful people to justify their actions. To address this flaw, you might introduce a second ethical precept: the "no harm" principle. This means avoiding unnecessary harm. So this is an example of how, by combining ethical principles and theories, we can try to develop a good theoretical foundation to our arguments for ethically advising and politically governing future developments of AI technologies up to and including the building of a superintelligent machine god (SMG).

To complete our understanding of how we can critically think about the ethics of AI, let's turn our consideration to the importance of rights and duties. You may be surprised to learn that the majority of the world does not know what it means to have a "right" to do something; or to have a "right" not to have something done to them.

Rights and Duties

Understanding how rights and duties function within any culture is not that complicated, but the majority of the world's population have little idea as to how they actually function. Basically, here's how they work: any time someone maintains that they have a right to do something, it entails that others have obligations or duties toward them in some respect. These obligations or duties come in two forms: active and passive. They match the two types of rights that can exist: positive and negative.

NEGATIVE RIGHTS

These are more basic and fundamental rights and involve the duty of non-interference by others. For example, if you claim to have a right to drink your coffee at the local coffee shop undisturbed, it means that neither I nor anyone else can disturb you while you do this. I cannot come into the coffee shop banging on drums and blowing trumpets and annoying you and other customers. So a negative right means that others have an obligation or duty not to interfere with your actions.

POSITIVE RIGHTS

These rights involve the duty of assistance. That is, if I claim I have a positive right to free access to specific buildings because I am wheelchair bound, it means that those who construct such buildings have a duty to make sure that I (and people in similar circumstances) can access the building (i.e., ramps, elevators, etc.). So, positive rights always entail duties of assistance. If you, or I, or anyone else in the world has a positive right to do anything, it means that others must, in some ways, assist us in attaining it. And that's it. Now you know more about how rights actually work in societies than most of the world's population.

When we consider a formal statement such as the Universal Declaration of Human Rights (UDHR),[31] we witness a global effort to determine which rights are the most important to all human inhabitants of this planet. It was drafted in Paris on December 10, 1948, by the United Nations (General Assembly resolution 217 A) as

> a common standard of achievements for all peoples and all nations. It sets out, for the first time, fundamental human rights to be universally protected and it has been translated into over 500 languages. The UDHR is widely recognized as having inspired, and paved the way for, the adoption of more than seventy human rights treaties, applied today on a permanent basis at global and regional levels (all containing references to it in their preambles).[32]

The UDHR contains thirty articles that discuss the rights humans have not only for life, liberty, and security of person, but also for freedom from slavery, punishment, discrimination, arbitrary arrest, and interference with one's privacy, family, home, correspondence, honor, and reputation. It also deals with freedom of thought and speech, education, work, culture, and so on. The majority of these rights are purely negative—for example, the right not to be subjected to arbitrary arrest, detention, or exile entails that others owe us a duty of noninterference. Other rights are positive—for example, the right to social security entails that one's country or state must assist those in need in some way.

Applied Ethics

In applying our critical thinking skills toward the construction of sound arguments for the safe and effective use of AI technologies, it would seem that a form of cafeteria- or buffet-style approach is necessary to bring in elements from various ethical theories, precepts, principles, and rights and duties mentioned above. In the field of ethics, this is known as "pluralism." The difficult part is deciding which of those mentioned above should or ought to be included and why. And then, we will have to determine the process in which such guidelines will be incorporated into the decision-making abilities of artificial narrow intelligence (ANI) as it emerges into artificial general intelligence (AGI).

To date, there have been many different proposals regarding the present and future use of artificial intelligence systems (AIS). At last count, there were no less than 167 separate proposals for ethical principles and guidelines.[33] I have scoured through the majority of these proposals and found that there is considerable overlap between them. In a thorough examination of the ethical guidelines of artificial intelligence, by far the most common themes tend to revolve around the following virtues, values, and principles:

1. Transparency/explainability
2. Value alignment
3. Avoid harm
4. Equality
5. Well-being
6. Autonomy
7. Privacy
8. Democracy
9. Inclusion
10. Shared responsibility
11. Sustainability
12. Security
13. Fairness
14. Respect
15. Accountability
16. Social benefit
17. Safety
18. Avoid creating or reinforcing unfair bias
19. Human control
20. Non-subversion
21. International cooperation for trustworthy AI
22. Open collaboration
23. Justice
24. Social harmony
25. Risk assessment
26. Contingency plans for failed AI activity

This list is not meant to be exhaustive but instead illustrative of the most commonly overlapping principles proposed throughout the world by numerous organizations in research, academia, industry, and world politics. These include but are certainly not limited to the following:

- Alan Turing Institute
- American Medical Association (AMA)
- Amnesty International
- Canadian government
- Chinese AI Alliance
- DeepMind
- Dubai: United Arab Emirates
- European Parliament
- European Union
- Future of Life Institute
- Indian government
- Israeli government
- G20
- Google
- IBM
- Institute of Electrical and Electronics Engineers (IEEE)
- Intel Corporation
- Japanese Society for AI
- Korean Ministry of Science
- Leaders of the G7
- Machine Intelligence Research Institute
- Microsoft
- Mozilla Foundation
- Organization for Economic Cooperation and Development (OECD)
- OpenAI
- Philips
- Pontifical Academy for Life
- Sony
- The Royal Society
- UK government
- United Nations Educational, Scientific, and Cultural Organization (UNESCO)
- United Nations General Assembly
- United Nations member states (193 total)
- U.S. government
- World Economic Forum

This partial list represents many of the very largest and most influential organizations and political powers in the world. What it demonstrates is that they all share a concern for the future safety and governance of AI. However, although these guidelines and principles are all agreed upon and shared across the vast majority of countries in the world, to what extent is anything being done? For we must now face the very difficult task of answering a number of very arduous and difficult questions:

1. How do we guarantee that these principles will be followed or practiced by *all* human parties involved as the development of AI technologies transforms our lives?

2. How do we police those who wish to willfully violate these principles?
3. Solving the alignment problem: How are we going to come up with a complete list of values, precepts, or virtues that will align, or allow us to control, or at least contain a superintelligent machine god when it passes from ANI to AGI and then to ASI?
4. Solving the consciousness problem: Should such an AGI/ASI develop sentience and consciousness, what rights—both ethical and legal—should it receive?

For the sake of brevity, let's call these the Big Four. Do you, dear reader, have any idea how incredibly difficult answering these questions is going to be? But as difficult as this will be, try we must. The future benefit of our world's inhabitants is depending on us—*all* of us—to get it right; there may be no second chance.

Global Use of Ethical Principles and Policing Detractors

1. Global Adherence to Ethical Principles

In early 2024, Volker Türk, the United Nations' high commissioner for human rights, visited Silicon Valley to meet with tech companies like OpenAI, Meta, and Google. He also spoke at various events at Stanford and Berkeley, and addressed representatives from Anthropic, Apple, Cisco, and Microsoft, to state a simple message: "Your products can do real harm and it's your job to make sure that they don't."[34] Türk believes that despite the great accomplishments that AI will be able to deliver to humankind, "they also hold enormous potential for addressing a range of societal ills [and] without effort and intent, these same technologies can act as powerful weapons of oppression."[35] In stressing the importance of complying with the existing UN Guiding Principles on Business and Human Rights,[36] Türk states: "you have already existing obligations and you need to apply them [because] business enterprises have the responsibility to respect human rights wherever they operate and whatever their size or industry [and] to know their actual or potential impacts [to] prevent and mitigate abuses."[37]

One of the current issues the world is facing right now is that these types of guiding principles are nonbinding. That is, they have no enforceable power—no teeth—as it were. This addresses questions 1 and 2 of the Big Four stated above. Although assuring that such ethical principles will be followed by all human parties involved in the development of AI technologies is going to be difficult, policing those who willfully violate these principles is going to be even more so. Obviously, within countries and nations, national, state, and local laws will preside over abuses with AI technologies that fall within their designated jurisdictions. But what happens when countries—or the leaders thereof—are guilty of violating international ethical codes of conduct and international law? What type of system can we create that is fair to the developers of the AI technologies while also protecting the rights of all those affected and punishing the violators? The world is definitely moving toward more detailed governance regarding such emerging problems—which we shall see in greater detail in chapter 5. But we are still in uncharted waters here. At this point in time, we still don't know how or when these principles will be accepted and practiced, if at all. Volker Türk says what gives him hope is what he refers to as the "silent majority" of people, who, he says, do "deeply care" about human rights, values, and dignity.[38] But he wants this silent majority to break their silence and speak up about what's going on in the world of AI: "I wish that the silent majority became a bit louder and were not silent, but actually, you know, overcome their fears, overcome the divisions and, and stand up for human rights."[39] In chapter 6, we shall consider how we, the members of the silent majority, can exercise our abilities to act toward a safer future of AI technologies. We can indeed become empowered with the knowledge and assurance that we have the ability to effect change and focus the direction and path of such emerging and powerful technologies.

For now, we have four very difficult questions to consider. And based on the main principles that academics, industry, and world leaders agree upon as guiding the governance of AI, we know now that, in moving forward, we will more than likely adopt a pluralistic[40] form of ethical structure in answering them.

> Whether our ethical practices are Western (Aristotelian, Kantian), Eastern (Shinto, Confucian), African (Ubuntu), or from a different tradition, by creating autonomous and intelligent systems that explicitly honor inalienable human rights and the beneficial values of their users, we can prioritize the increase of human well-being as our metric for progress in the algorithmic age.[41]

And so it would appear that a form of pluralistic principlism will be used in the establishment of agreed-upon ethical precepts, values, and theories.

2. How Do We Police the Cheaters?

To consider how we might proceed in guaranteeing that our agreed-upon ethical principles will be followed or practiced by all human parties involved and how we might police those who wish to willfully violate these principles, let's consider some actual cases occurring in real time.

Let's turn our attention to a pressing ethical concern that has arisen with self-driving or autonomous vehicles. We need to think very carefully about the so-called "choices" these cars might have to make in dangerous situations. For instance, if an autonomous vehicle faces an unavoidable collision, what moral priorities should it follow? Should it prioritize crashing into other vehicles instead of pedestrians? Or a building? Should it prioritize sparing humans over pets? Prioritize certain genders or age groups? Consider income status or whether someone is following the law? Choose utilitarianism over deontological guided principles? These questions lead to challenging ethical dilemmas not unlike our trolley problem. Let's focus on one scenario: if a self-driving car has no choice but to crash on a busy urban street, and it's going to hit either a child or an elderly person, who would you rather be hit? Surprisingly, the answer depends on cultural differences. In North America, self-driving cars are designed to prioritize saving young lives. However, in Asian countries like China, Japan, and South Korea, there's a greater emphasis on saving the elderly.[42] So which or whose approach is right? In developing a sound argument for or against these types of preferences, some might point to the ethical concept of cultural relativity. If it's just a matter of culturally relative differences, like "When in Rome, do as the Romans do," should we have the option to change the car's priorities when we use it in different countries? For example, if someone from Japan is traveling in the United States, should they be allowed to change the car's moral rules to match their personal cultural preferences in case of an accident? And vice versa, should Americans be allowed to do the same when in Japan? If I'm an animal rights advocate, should I be allowed to change the settings to avoid hitting animals over humans? This is just one example of an ethical dilemma that gives us a lot to think about, discuss, and figure out in regards to how, exactly, we will choose our ethical principles, values, or theories as AI technologies continue to advance.

Let's now consider the extent to which AI technologies should be used to create artistic works of art either by dead artists or by living artists without their permission. Two examples come to mind: the AI-generated song "Heart on My Sleeve," with vocals made to sound like Drake and

The Weeknd; and the use of a "stunt voice" of Bette Midler in a television commercial.

Since neither Drake nor The Weeknd sang, participated in, or sanctioned the development of the song, how do they get compensated, if at all? Or stop others from using their material with AI technologies? In the United States, copyright laws have weighed in on somewhat similar cases. Back in 1988, a Ford car commercial tried to hire a singer who sounded just like Bette Midler for one of its ads. Midler successfully sued because, in California, the courts decided that she possessed common law rights to her voice and allowed her to advance her lawsuit. When the AI-generated Drake/Weeknd work surfaced,

> Universal Music Group requested "Heart on My Sleeve" be removed from the streaming platforms using copyright take-down requests. The song contains a short audio sample of the well-known tag of producer Metro Boomin, which does have copyright protection. But without that tag, UMG likely wouldn't have any direct copyright claims to an otherwise original song with original lyrics. Courts have held that voice or musical style can't receive copyright protection.[43]

It's interesting to note that, because a human being (Metro Boomin) produced audio material within the song, it contains copyright protections. But if it had been produced purely by AI, it might not. This raises another interesting issue, which considers to what degree, if any, do AI creations break existing copyright laws? Some have argued that, since LLMs are trained on material that has already been copyrighted, it follows that they are in violation of such laws when they produce "new material."

> It remains unresolved whether AI programs—like those that can produce voice imitations or new melodies—violate copyright law by training on millions of copyrighted songs and sound recordings without a license. Companies behind generative AI art programs are facing different lawsuits brought by artists, open source coders, and Getty Images over machine-learning algorithms that swept up their content.[44]

One might infer that if images are to be protected under copyright laws, it would follow that sounds, vocalizations, melodies, recordings, and such would be as well.

This leads us to consider the George Carlin case again.[45] Since George Carlin died in 2008, we must consider to what degree has the use of his style been a violation of his rights—both ethical and legal? As I mentioned earlier, his daughter, Kelly, thinks so. Full confession: Kelly and I have known each other for quite a few years now. I interviewed her on my podcast, and we've spoken at great length about her father's life.[46] Kelly proceeded with litigation:

> Carlin's daughter, Kelly Carlin, said in a statement that the work is "a poorly-executed facsimile cobbled together by unscrupulous individuals to capitalize on the extraordinary goodwill my father established with his adoring fanbase." The Carlin estate and its executor, Jerold Hamza, are named as plaintiffs in the suit, which alleges violations of Carlin's right of publicity and copyright. The named defendants are Dudesy and podcast hosts Will Sasso and Chad Kultgen. "None of the Defendants had permission to use Carlin's likeness for the AI-generated 'George Carlin Special,' nor did they have a license to use any of the late comedian's copyrighted materials," the lawsuit says. The defendants have not filed a response to the lawsuit and it was not clear whether they have retained an attorney. They could not immediately be reached for comment. At the beginning of the special posted on YouTube on Jan. 9, [2024] a voiceover identifying itself as the AI engine used by Dudesy says it listened to the comic's 50 years of material and "did my best to imitate his voice, cadence and attitude as well as the subject matter I think would have interested him today." The plaintiffs say if that was in fact how it was created—and some listeners have doubted its stated origins—it means Carlin's copyright was violated.[47]

As mentioned earlier, we have since learned that the entire thing appears to be a bit of a hoax—that is, the stand-up routine was created by humans—YouTube entertainers Will Sasso and Chad Kultgen—which was then made to look like it was AI generated. It looks as though it is nothing more than a fairly clever *Victor/Victoria*[48] theme; but it is still ethically problematic nonetheless. Both Sasso and Kultgen are clearly in violation of George Carlin's rights to autonomy and dignity. His persona is being used as a means to an end for the purposes of their personal gain. This has denied George Carlin (or, in his absence, his daughter, Kelly, and his estate) any power of control or designation.

The recent abuse of George Carlin's ethical and legal rights raises some very important questions, not the least of which is this: Who owns the likenesses and imitative control of any deceased person's accomplishments? And how far back in time would this extend? For example, if I wanted some AI program to create a likeness of Socrates based on all recorded information about him, can anyone sue me? If I created a lifelike chatbot or robot of Socrates that knew all of the writings of Plato and all other relevant teachings during his time, if there are no legal repercussions, are there any ethical issues? Are there licensing windows after which the public has access to use an artist's material in various ways? What are the moral and legal statutes of limitations on using AI to recreate dead actors, musicians, comedians, entertainers, and politicians, among others? This is something we need to start thinking about sooner than later. Because it won't be long before you can rent or buy yourself a Queen Victoria or Martin Luther King likeness. We already use Puccini, Beethoven, and Tchaikovsky's amazing works to sell yogurt, cars, and chocolate. So why not take it a step or two further and let AI generate us up some historical simulations to entertain us?

And as a final example, let's consider what AI technologies are going to do to the employment landscape. As we saw in chapters 2 and 3, jobs will be created by AI, and jobs will be replaced by AI. It would seem that the best argument moving forward would be to embrace all that AI can do for us in a changing work environment while, at the same time, balance the amount of jobs lost with either job or income replacement. Ideally, we want the transition period of AI development to be as smooth and fair as possible. So, principles of fairness, benefit, respect, justice, and the avoidance of harm appear to be among the top priorities through such a transitional phase. In terms of ethical theories, we would want the greatest good for the greatest number of people while observing the virtues of compassion and empathy. In other words, we want people to be able to either work or receive compensation for having lost their jobs due to AI advancements. What we don't want is millions of people with nothing to do who are very angry about the fact that they've been replaced by a machine. To avoid this type of potential resentment and reprisal, it is very important to stay ahead of the ways in which AI will and is changing the employment landscape. We want our governments and industry leaders to be Promethean in looking forward and anticipating these changes. And we also want to be able to ensure that legally, the rights of people will not be violated. Proper plans to allow for effective monitoring and the policing of these changing environments will also be of vital importance. In these ways, we can all be more proactive rather than reactive to the changes that are coming. Let's now consider questions 3 and 4 of the Big Four: the alignment problem and the problem of consciousness.

3. Trying to Solve the Alignment Problem
I'm not sure how to tell you this, dear reader, but solving the alignment problem appears to be extremely difficult to accomplish. Currently, there are several major organizations in the world attempting to figure out how

we might program a superintelligent machine god (SMG) so that it stays aligned to our values and does *only* what we want it to do. As you might imagine, there are several schools of thought on this issue—there are those who believe this can be accomplished and those who believe it will never be accomplished. Those who are trying to solve this problem include, but by no means are limited to, OpenAI, Machine Intelligence Research Institute (MIRI), Center for Human-Compatible AI (CHAI), Future of Life Institute (FLI), Berkeley Existential Risk Initiative (BERI), the Partnership on AI (PAI), the Alignment Research Center (ARC, now METR), and Center for AI Safety (CAIS). These organizations, along with a considerable number of academics, researchers, philosophers, and scientists, are devoting their efforts right now to one, singular purpose:

To solve the alignment problem.

Although there are plenty of other AI risks, harms, and problems that we must direct our attention toward and solve, successfully aligning an SMG is one of the biggest and most important problems to solve in AI right now. For, if we can solve it, then perhaps we can control or contain an SMG to give us all of the good stuff we ask of it while, at the same time, prevent it from either accidentally or intentionally harming or killing us. But this, of course, is no guarantee that some will choose to use powerful forms of AI for nefarious or harmful reasons. Alignment is one problem; getting everyone to get along and not abuse the powers of an SMG are quite another.

But what human values do we program into such a god? It seems life is not without irony. For millennia, humans have believed it was various incarnations of gods who bestowed upon humans ethical rules for our behavior—rewards if we listen and abide; punishments, if we do not. And now, it is us, the humans, who have to figure out how to make a god behave itself. So then, how do we do this? One particular group, AI Alignment Proposals, presents research dealing with many different facets of the problem:

> Aligning advanced artificial intelligence to human values requires concerted research across fields from machine learning to ethics. Interactive learning, imitation learning, crowdsourcing, oversight, transparency, and conservatism each contribute partial solutions. Employing an integrated approach combining these complementary techniques offers a robust means of cultivating AI systems that build upon humanity's moral wisdom rather than subverting it. If guided by proactive compassion and creativity,

we can harness AI to profound benefit while aligning its goals to our highest shared values through this process of cooperative engagement. With diligence and care, artificial intelligence can become our ally in realizing both enlightened ideals and pragmatic progress for all.[49]

Although we find references to limiting bias, incorporating transparency, and validating methods to determine if such a superintelligent system is aligned with "our highest shared values," there is no explicit mention of what specific values in particular are being used to train and control such an entity.

There are those who believe we can align such a god to our values by tinkering with it in just the right way. Scholars like Max Tegmark (MIT), Steve Omohundro (CEO of Beneficial AI Research), Anthony Aguirre (executive director and board secretary of Future of Life Institute), Jan Leike (head of OpenAI's Superalignment Project), and others believe the best way forward in solving the alignment problem is to "box" or contain the god; or limit its capacity for computing power; or "close the gate" on it; or simply design safer systems from the start with checks and balances. For Anthony Aguirre, the best way to control a superintelligent machine god is to contain it:

> This essay is an extended argument for why we should not, in the next few years, irrevocably open this gate: we should not train neural networks better than nearly everyone at a wide range of intellectual tasks, let alone ones better than the very best human experts or even all of human civilization. Instead, we should set an indefinite hard limit on the total computation employed in training an individual complete neural network, a bound on how fast such a neural network runs, and likely other limits.[50]

Aguirre mentions that there have been other forms of technology where we have consciously decided to either not develop or cease advancing because of the potential for harm, for example, human cloning, human germ-line engineering, eugenics, and advanced bioweapons, among others.[51] In much the same manner, so too should we consider AGI systems to be "outside the gate" in our abilities to control or contain them. Aguirre poses his extended argument as follows:

1. We are at the threshold of creating expert-competitive and superhuman [AGI] systems in a time that could be as short as a few years.

2. Such "outside the Gate" systems pose profound risks to humanity, ranging from, at minimum, huge disruption of society to at maximum permanent human disempowerment or extinction.

3. AI has enormous potential benefits. However, humanity can reap nearly all of the benefits we really want from AI with systems inside the Gate, and we can do so with safer and more transparent architectures.

4. Many of the purported benefits of [AGI] are also double-edged technologies with large risk. If there are benefits that can *only* be realized with superhuman systems, we can always choose, as a species, to develop them later once we judge them to be sufficiently—and preferably provably—safe. Once we develop them there is very unlikely to be any going back.

5. Systems inside the Gate will still be very disruptive and pose a large array of risks—but these risks are potentially manageable with good governance.

6. Finally, we not only should but can implement a "Gate closure": although the required effort and global coordination will be difficult, there are dynamics and technical solutions that make this much more viable than it might seem.[52]

This is an optimistic proposal that sounds both viable and hopeful. But it hinges on one very difficult aspect of the human condition: the drive to see "what will happen next." Aguirre assumes or begs the question that humans will be cooperative in this regard. Personally, I applaud his proposal. I think it is quite sound. And I sincerely hope it works, if adopted. However, it requires universal or global cooperation in order to work. And I'm not convinced we're capable of this—at least not all countries (or Big Tech companies). Perhaps the majority of countries and nation-states throughout the world will comply. But what about those who decide to flout such agreements? This is why the world needs to get very Hobbesian very quickly. We need to establish global agreements with all countries to abide by such measures to assure mutual benefit and avoid mutual harm or destruction. It's literally a hang together or hang separately scenario. And this requires something along the lines of a global accord. We shall consider this more deeply in the next chapter. But for now, it seems overwhelmingly obvious that the majority of the countries and nation-states in the world agree to the shared principles and values mentioned above. Now, we just need to get all world leaders to sign a contract—an accord—and agree to abide by it, and expect to be punished if they violate it. If part of the accord turns out to be that we close the gate on AGI for now, so be it. Or perhaps we can "box" it or confine it in a way that limits its abilities.[53] But what if this is not possible? Are there other plans? There's actually quite a few. But we'll just look at a couple more for now.

In 2023, Max Tegmark and Steve Omohundro produced an influential paper titled "Provably Safe Systems: The Only Path to Controllable AGI."[54] In this paper, they argue the following:

1. AGI is imminent.
2. AGI is an existential risk.
3. AI alignment is insufficient.
4. We need a security mindset.
5. We need provable safe systems.
6. AI theorem proving is critical and imminent.
7. We need cryptographically secure hardware.
8. This is the (only) way.
9. Success will lead to human thriving.[55]

But what comprises a "provably safe system"? Tegmark and Omohundro believe that it is possible to create systems with hardware that will be provably compliant with a set standard within all levels of computer development. They discuss the potential for developing proof-carrying code (PCC) through the use of limited AI, which will then monitor all provably compliant hardware (PCH) and operations for compliance according to risk-commensurate specifications. In this way, they believe that it will be possible to control AGI through a compliance-measuring tactic.

> In summary, we've described a vision for how humans can control AGI and superintelligence, where the only AGI that ever gets deployed consists of proof-carrying code. AI is allowed to write the code and proof, but not the proof-checker. The knowledge and algorithms used by the code can be AI-discovered, extracted either through self-explaining AI or with mechanistic interpretability tools. We've argued that this vision isn't merely plausible, but that it may be the only guaranteed solution to the control problem: no matter how superintelligent an AI is, it can't do what is provably impossible. Losing control over superintelligent AI would arguably be humanity's worst mistake ever, so proponents of alternative approaches to AGI safety owe a detailed explanation of how they guarantee success. For example, passing evaluations that screen for known risks are a necessary but not sufficient condition for safety. Absence of evidence of risk isn't evidence of its absence. Mathematical proofs are![56]

When I hosted Steve Omohundro on my podcast, *All Thinks Considered*, I asked him when he thought AGI might emerge. He said:

> Amongst the people I talk to, it's obvious AGI is coming. Exactly when, some are saying two three years. Metaculus [predicts by] 2032; so, eight years. Jensen Huang, the head of NVIDIA . . . estimates that is five years for full AGI. Well what happens once you have full AGI? Well, it's as good as humans at anything humans do, in particular, writing AI algorithms and creating new things. So these systems, presumably, will be able to improve themselves. And in fact, Jensen says that like the H100, the powerful chip that they make that is driving all this AI stuff, that I think they're like 40,000 circuits in that chip that were designed by AI. So already we have AI improving the substrate on top of which AI sits. And so some people think that process will speed up dramatically . . . once there's an AGI that can do something, you can copy that program a million times. So now you have an Einstein AGI making copies of itself and now you have a million Einstein AGIs. If they can talk to one another and relate, you have a company of a million geniuses working on something. Some people believe that the progress at that point will just really ramp up and become much more quicker.[57]

So we may see a network of distributed intelligences combining to create a greater entity. When I asked Steve about how we might control these very powerful forms of AI, he stated that we can go one of two ways:

> I think humanity is at a fork in the road, where if we do things right, we will suddenly have these huge intelligences available to solve basically all the problems of humanity: solve climate change, solve poverty, solve health issues, cure cancer . . . all that stuff; we're just right on the verge of that. On the other hand, if we don't choose our goals and things properly, these AIs will be great at warfare. They'll be great at, you know, breaking into banks and stealing money. They'll be great at manipulating people to get them to do whatever they want. And so we could end up easily in a dystopia where the goals that the AI is pursuing aren't aligned with what most humans care about. And there are many forms of dystopia. And some of them end up with them saying, "we don't need these pesky humans; yeah, thank you, humans, for building us, but we're done with you." And the humans go extinct. And there are explicit stories about how that happens. Nukes are a big, big thing to think about . . . and why would the AIs want to get control of the nukes? Well, if they're in control of the nukes, they can get anything they want; they say, "Gee, you know, we really would like much bigger data centers, and we'd really like to spread out over the entire Sahara Desert. And if you don't do that, well, we're just gonna launch these nukes."[58]

But the question still remains: How are we going to control a superintelligent machine god? What controls can we put in place that will disallow an SMG from going rogue or misaligning from our ethical guardrails? Omohundro says: "What Max [Tegmark] and I are doing is we're thinking, okay, what are the attacks that an AGI could put onto some system, say, hack into a nuclear launching facility or a biohazard lab or something like that? And how could we harden that so that even the most intelligent, most adversarial, most horrible AGI in the world still couldn't cause something that would be, you know, existential risks for humans?"[59] The way in which Omohundro and Tegmark propose to control an SMG is through two means: the laws of physics and mathematical theorem checking:

> [Let's say] we have this new intelligence, [and] it's not aligned, and it has its own goals. How on earth can we possibly control it? So Max and my insight was that there are two things that, no matter how intelligent it is, it cannot break. And that's the laws of physics and mathematical proof. And so physics has these core fundamental things, conservation of energy, you know, the speed of light; no AI, no matter how intelligent it is, is going to be able to go faster than the speed of light [and] it's not going to be able to create matter out of nothing. And so we get constraints on what possible actions that could take from that.[60]

And just as the laws of physics cannot be usurped or broken by any form of artificial intelligence, so too does Omohundro believe the same holds true for a type of foundational mathematical modeling:

> In the early part of the nineteenth century, mathematicians developed a foundation for all of mathematics . . . the best is called Zermelo-Fraenkel set theory, where you can encode any mathematical statement. And then you can precisely check step by step if an argument is correct, and get absolute truth out of it. And the beauty of that is that we can encode all of the physical laws in that setup; we can encode engineering, we can encode programming languages, computers. . . . And so we have a foundation where we can encode all of the things in the world that we care about, and we can reason about what their behavior could be, or will be in a way that we can check mechanically. And that is mathematical theorem proof and theorem checking. And so in our proposal, those two pieces, the laws of physics and this theorem checking, show that even the most powerful, most intelligent AGI can't prove a false theorem. They can never show that two plus two equals five and so on. That gives us weak, not so smart humans, absolute control.[61]

If you're having trouble understanding exactly what and how this type of control measure is going to tame an SMG, you're not alone. Basically what Omohundro and Tegmark are suggesting is that we will be able to use the immutable laws of physics and mathematical theorem checking as "gatekeepers" that will stop any SMG from doing things we don't want it to do.

> So we have this crazy AGI doing who knows what. And it may, through part of its thinking, want to do some bad thing, like launch nukes or whatever. We have a gatekeeper there, that is a physical device, a physical object in which we have designed it and built it so that by the laws of physics, it's impossible to, you know, break the cryptographic code that it uses to make the thing happen. And so that's the source of state safety. And if we can make that little teeny device not unnecessarily expensive or hard, then we can rebuild the nature of interactions in our society, out of these components that are themselves unbreakable. And that have rules which are encoded in this mathematical logic that get checked automatically. And interestingly, [it will stop] the AGI [from doing] bad things.[62]

So there will be controls over potentially harmful and misaligned AGIs; but what about people who might try to override controls and use such powerful forms of AI to further their evil ends? Omohundro says:

> We need to make sure no humans can do existentially bad things. And so that's going to be a shift in thinking, you know, I'm sure that's gonna add some challenge but if we can do that, then we argue that it's actually not that difficult to create gatekeepers around the most dangerous things in the world. And that would at least keep us away from the existential risks.[63]

I was impressed with Omohundro's confidence that his and Tegmark's system for limiting existential risks from AI will work. It will be interesting to see how the next steps in AI governance will advance with such proposed solutions. In other words, even if Omohundro and Tegmark's provably safe systems approach works, will the right people in power listen? As with all scientific endeavors, we will need to confirm a proof of concept that such a solution is not only viable but applicable.

In the meantime, others are proposing different approaches to solving the alignment problem. For Jan Leike, who heads OpenAI's Superalignment Project: "What we want to do with alignment is we want to figure out how to make models that follow human intent and do what humans want—in particular, in situations where humans might not exactly know what they want."[64] When asked whether or not he believed that ChatGPT was "aligned," Leike said:

> I wouldn't say ChatGPT is aligned. I think alignment is not binary, like something is aligned or not. I think of it as a spectrum between systems that are very misaligned and systems that are fully aligned. And [with ChatGPT] we are somewhere in the middle where it's clearly helpful a lot of the time. But it's also still misaligned in some important ways. You can jailbreak it, and it hallucinates. And sometimes it's biased in ways that we don't like. And so on and so on. There's still a lot to do.[65]

For the record, a "jailbreak" is the ability by a human to prompt a large language model like ChatGPT to provide information that bypasses its content moderation guidelines. In other words, "jailbreaking" is a hack technique that fools systems like ChatGPT to provide dangerous information it is not supposed to provide (e.g., harmful information involving the creation of weaponry, causing harm to animals and humans, etc.). "Hallucinations," on the other hand, refer to text information that is simply false or extremely exaggerated. It's as though ChatGPT is making stuff up when it responds to a prompt. One of the most famous cases of this involved a New York attorney named Steven A. Schwartz:

> A New York lawyer cited fake cases generated by ChatGPT in a legal brief filed in federal court and may face sanctions as a result, according to news reports. . . . The incident involving OpenAI's chatbot took place in a personal injury lawsuit. . . . Judge P. Kevin Castel wrote in an early May order regarding the plaintiff's filing that "six of the submitted cases appear to be bogus judicial decisions with bogus quotes and bogus internal citations." He called it "an unprecedented circumstance."[66]

Do your homework, kids. And always check your sources. I'm sure Mr. Schwartz will never forget this lesson.

Anyhow, in reference to the alignment problem, Leike addresses three different types:

> ChatGPT can hallucinate and give biased responses. So that's one level of misalignment. Another level is something that tells you how to make a bioweapon. And then, the third level is a superintelligent AI that decides to wipe out humanity.[67]

Leike and his team are focused primarily on the third, or most severe form of misalignment: existential risk (x-risk). But what strategies are currently being deployed in addressing this type of misalignment?

Basically, if you look at how systems are being aligned today, which is using reinforcement learning from human feedback (RLHF)—on a high level, the way it works is you have the system do a bunch of things, say, write a bunch of different responses to whatever prompt the user puts into ChatGPT, and then you ask a human which one is best. But this assumes that the human knows exactly how the task works and what the intent was and what a good answer looks like. And that's true for the most part today, but as systems get more capable, they also are able to do harder tasks. And harder tasks will be more difficult to evaluate.[68]

Figure 4.11.   The Shoggoth. *Created for the author. All rights to the author.*

The use of reinforcement learning from human feedback (RLHF) to better align LLMs with human values has jokingly been referred to in the AI biz as a Shoggoth (figure 4.11).

> Shoggoths are fictional creatures, introduced by the science fiction author H. P. Lovecraft in his 1936 novella "At the Mountains of Madness." In Lovecraft's telling, Shoggoths were massive, bloblike monsters made out of iridescent black goo, covered in tentacles and eyes. Shoggoths landed in the AI world in December, a month after ChatGPT's release, when a Twitter user, @TetraspaceWest, replied to a tweet about GPT-3 (an OpenAI language model that was ChatGPT's predecessor) with an image of two hand-drawn Shoggoths—the first labeled "GPT-3" and the second labeled "GPT-3 + RLHF." The second Shoggoth had, perched on one of its tentacles, a smiley-face mask.[69]

Simply put, the joke is that, to make sure AI language models don't act scary or dangerous, AI companies teach them to be polite and harmless. They use a method called "reinforcement learning from human feedback" (RLHF), where humans rate chatbot responses, and the AI apparently "learns" from those scores.

> Most AI researchers agree that models trained using RLHF are better behaved than models without it. But some argue that fine-tuning a language model this way doesn't actually make the underlying model less weird and inscrutable. In their view, it's just a flimsy, friendly mask that obscures the mysterious beast underneath. . . . Comparing an AI language model to a Shoggoth . . . wasn't necessarily implying that it was evil or sentient, just that its true nature might be unknowable.[70]

For most, the Shoggoth represents the unknown, unpredictable, and potentially dangerous qualities and characteristics of AI.

> Today, if you hear mentions of the Shoggoth in the AI community, it may be a wink at the strangeness of these systems—the black-box nature of their processes, the way they seem to defy human logic. Or maybe it's an in-joke, visual shorthand for powerful AI systems that seem suspiciously nice. If it's an AI safety researcher talking about the Shoggoth, maybe that person is passionate about preventing AI systems from displaying their true, Shoggoth-like nature.[71]

It's always interesting to see, in real time, what members of the AI community are saying about current developments. Referring to AI as a hideous alien-like monster that seems friendly but may be deadly is quite accurate because we really don't know what such systems are and will be capable of—no matter how many smiley faces we stick on them:

> In any case, the Shoggoth is a potent metaphor that encapsulates one of the most bizarre facts about the AI world, which is that many of the people working on this technology are somewhat mystified by their own creations. They don't fully understand the inner workings of AI language models, how they acquire new abilities or why they behave unpredictably at times. They aren't totally sure if AI is going to be net-good or net-bad for the world . . . what's happening in AI today feels, to some of its participants, more like an act of summoning than a software development process. They are creating the blobby, alien Shoggoths, making them bigger and more powerful, and hoping that there are enough smiley faces to cover the scary parts.[72]

So, what would happen if we simply replaced human reinforcement learning with particular narrower forms of AI? Some believe this might speed up and more accurately create a safer environment. Leike is not alone in believing that we may be able to use narrower forms of AI to help us in keeping AGI in line with our moral guidelines.

> The idea behind scalable oversight is to figure out how to use AI to assist human evaluation. And if you can figure out how to do that well, then human evaluation or assisted human evaluation will get better as the models get more capable, right? For example, we could train a model to write critiques of the work product. If you have a critique model that points out bugs in the code, even if you wouldn't have found a bug, you can much more easily go check that there was a bug, and then you can give more effective oversight.[73]

As we saw earlier, the difficulty in knowing what is going on inside an AI system is known as the "black box problem." As noted earlier, this refers to the considerable difficulty of understanding what an AI system is doing once it has been granted considerable power plus autonomous activity:

> Some researchers . . . think what ultimately may be necessary is what AI researchers call "interpretability"—a deep understanding of how exactly models produce their outputs. One of the problems with machine-learning models is that they are "black boxes." A conventional program is designed in a human's head before being committed to code. In principle, at least, that designer can explain what the machine is supposed to be doing. But machine-learning models program themselves. What they

come up with is often incomprehensible to humans. Progress has been made on very small models using techniques like "mechanistic interpretability." This involves reverse-engineering AI models, or trying to map individual parts of a model to specific patterns in its training data, a bit like neuroscientists prodding living brains to work out which bits seem to be involved in vision, say, or memory. The problem is this method becomes exponentially harder with bigger models.[74]

Since we have less and less of an idea of what's going on inside the black box brain of a superintelligent machine god, another problem poses itself: deception. How will we be able to tell if this very new and powerful form of AI is telling us the truth about its behavior or intent?

> Evaluating these really high-level things is difficult, right? And usually, when we do evaluations, we look at behavior on specific tasks. And you can pick the task of: Tell me what your goal is. And then the model might say, "Well, I really care about human flourishing." But then how do you know it actually does, and it didn't just lie to you? And that's part of what makes this challenging. I think in some ways, behavior is what's going to matter at the end of the day. If you have a model that always behaves the way it should, but you don't know what it thinks, that could still be fine. But what we'd really ideally want is we would want to look inside the model and see what's actually going on. And we are working on this kind of stuff, but it's still early days. And especially for the really big models, it's really hard to do anything that is nontrivial.[75]

This is often referred to as the "interpretation problem." It involves the very difficult task of peering into the black boxes to accurately identify how such advanced AI systems "thought" and "acted" the way they did. These systems process mountains of data and make thousands of decisions; but it's not always clear how they weigh different factors or why they make a specific choice. The interpretability problem is primarily concerned with making these advanced AI systems more transparent and understandable to humans. Abiding by these principles is universally accepted in the AI fields because we all want to know with certainty why an AI system thinks a certain way so we can trust it more and know that it's making fair and logical decisions.

So the race is on to build a superintelligent machine god that not only might be extraordinarily good at keeping its thinking secrets from us but also may develop the capacity to deceive us. This leads us to consider if we might be able to avoid such a state by assuring such systems align to our values. To find out, Leike and his team are intentionally building

ChatGPT models that deliberately attempt to deceive them, known as "red-teaming"—or the deliberate attempt to hack an AI system to make it do things it's not supposed to do. Leike believes:

> If we deliberately make deceptive models, A, we learn about how hard it is [to make them] or how close they are to arising naturally; and B, we then have these pairs of models. Here's the original ChatGPT, which we think is not deceptive, and then you have a separate model that behaves basically the same as ChatGPT on all the ChatGPT prompts, but we know it has this ulterior motive [to lie] because we trained it to be that way. And then you can compare them and say, okay, how can we tell the difference? And that would teach us a lot about how we would actually catch these things arising in the world.[76]

In case you're still wondering what "red-teaming" is, consider this definition:

> Another approach, borrowed from war-gaming, is called "red-teaming." OpenAI worked with the Alignment Research Centre (ARC), a non-profit, to put its model through a battery of tests. The red-teamer's job was to "attack" the model by getting it to do something it should not, in the hope of anticipating mischief in the real world.[77]

By his own reckoning, Leike anticipates that such a "red-teaming" method may potentially cause the very AI system to learn how to become a better deceiver, known in the AI biz as "self-exfiltration." Leike believes "we just [need] to train them to be deceptive in deliberately benign ways where instead of actually self-exfiltrating you just make it reach some much more mundane honeypot. You have to put really great care into the sandbox that you put around this experiment."[78] I love his use of the term "just," as in "you *just* make it." I have always told my kids how problematic the word "just" can be. For it implies that there is simply a bare minimum that is required to make everything OK. But again, we're in uncharted waters here. Maybe it can work; maybe not. But there seems to be an awful lot riding on these safeguards. Let's "just" hope they work.

Leike believes there are three ways an AGI machine god could "self-exfiltrate":

> A, the model could persuade an OpenAI staff member to help it . . . B, it could use social engineering and pretend to be someone else to trick someone to do this. Or C, it could figure out how to break the technical measures that we put in place to secure the model. So our goal here would be to understand exactly where the model's capabilities are on each of these

tasks, and to try to make a scaling law and extrapolate where they could be with the next generation. The answer for the models today is they're not really good at this. Ideally, you want to have the answer for how good they will be before you train the next model. And then you have to adjust your security measures accordingly. . . . This is why understanding the model's risk for self-exfiltration is really important, because it gives us a sense for how far along our other alignment techniques have to be in order to make sure the model doesn't pose a risk to the world. [79]

Leike proposes the use of AI to determine the level of alignment of next-generation AI; the type of powerful AI that precedes the evolution to AGI.

Basically, the idea is if you manage to make, let's say, a slightly super-human AI sufficiently aligned, and we can trust its work on alignment research—then it would be more capable than us at doing this research, and also aligned enough that we can trust its work product. Now we've essentially already won because we have ways to do alignment research faster and better than we ever could have done ourselves. And at the same time, that goal seems a lot more achievable than trying to figure out how to actually align superintelligence ourselves.[80]

Yes, Leike may be correct that such alignment AI bots might be able to help us keep the superintelligent machine god in line but his assumptions are showing. Leike begs the question when he says "and we can trust its work on alignment research" by assuming that we can know this with certainty. It raises the problem of the potential for deception again and whether or not and to what degree we might ever be able to rule this out.

We have looked at some of the current developments in the drive to solve the alignment problem. As you will notice by now, it is a very difficult problem to solve with both humans and superintelligent machine gods alike. Now that we have considered three of the Big Four questions, let's consider the final, and perhaps most interesting of all.

4. The Consciousness Problem

At this point in history, when we are faced with our "Oppenheimer moment," it's time we consider something very important in our race to build such a god: what happens if it becomes conscious? What if it evolves to the point whereby it knows it's "alive" in some respects? To what extent will it comply with its moral guardrails that we humans have put in place to make it behave if it becomes consciously aware of its own existence? The answer is: we don't know. And this is a really big "known unknown." It is at this particular point that we might want to ask ourselves these questions: What is consciousness? What do we mean when we say someone or something has consciousness? What does this involve? Perhaps we should consider how humans developed this most wondrous of features if we are to begin to understand how a machine god might do the same.

I have researched, written, and lectured extensively on cognitive evolution and the coevolution of human consciousness.[81] And although entire courses, series, conferences, journals, and books are devoted to the pursuit of defining consciousness, I will present a composite of defining features that various scholars and schools of thought generally agree are essential for the existence of human consciousness:

1. We, humans, are *perceptually aware* of our environments.
2. We are *self-aware*.
3. We can *modify our ideas and memes according to a value system* in the attempt to increase the likelihood of our survival.
4. We developed a strong sense of *mortality*.
5. We have a fully developed *theory of mind*.

To elaborate, human consciousness involves not only that we are *perceptually aware* of our environment—for most species are aware (from amoebas to zebras)—but also that we are unique in the respect that we are also aware *that* we are aware of ourselves, our environment, and the relationship between the two. In other words, we are *self-aware* and are cognizant that we are both distinct from and a part of (an) ecology. This is what some philosophers refer to as a type of second order or "meta" perception often referred to as "meta awareness." We also have the capacity to recognize the relationship between our beliefs and our ecology. We can, through trial and error, *modify our ideas* in the attempt to increase the likelihood of our survival (e.g., better nutrition, medicine, housing, etc.). And we can acknowledge benefit or harm relative to a *value system*, which measures their success rate as they are incorporated into survival strategies. Another extremely important factor in the evolution of consciousness is that we have a strong sense of *mortality*. As depicted in the eventual customs and burial rituals of our ancestors, we know, by analogy, that we are going to die. This paints a very clear conceptual picture about our own lives. We are a comparative species. We constantly make associations (as do many other species). But no other species nears our capacity for symbolic representation or material manipulation with regard to these associations and comparisons. As a deeply comparative species, somewhere in our evolutionary past, our ancestors used analogical reasoning to realize that their lives—like those of their dead kin—were going to end. The "Oh s#★t, that's going to be me someday!" realization led to the first caveman angst or existential *Homo sapien*, which led to burials and symbolic representations of the afterlife. When you realize that you're going to die someday, that really brings perspective to one's existence. It singles us out and defines us in ways that, up until that point, we were quite incapable. In other words, in our evolutionary past, the archaeological records indicate that we did not understand our own finitude (i.e., that we were going to die). But we gradually came to this realization. Knowing that our existence will cease someday dramatically increases our realization of our own existence. It provides a very real and powerful dichotomy separating or categorizing the living from the dead. Understanding this relationship as it relates to our own existence raises our consciousness and awareness about ourselves and the world in which we live. And finally, with our keenness for cataloging comparisons, we have become acutely aware of the relationship among ourselves, our kin, and our environments. This has led to the development of a *theory of mind*. That is, we developed a very good capacity for not only seeing the world from the vantage of others but also making associations about the behavior of others based on that capacity.[82] This has given rise to an increased ability for understanding social standing, for making inferences about the thoughts of others, for spotting deception, and for emotional feelings such as empathy. Both Nicholas Humphrey and Richard Leakey believe that human brain encephalization led to the increased ability to monitor one's own behavior within a group.

> If, for example, individuals were able to monitor their own behavior, rather than merely operate as computerlike automatons, then they would develop a heuristic sense of what to do under certain circumstances. By extrapolation, they might then be able to predict the behavior of others under the same circumstances. . . . Once consciousness was established, there was no going back, for individuals less well endowed would be at a disadvantage. Similarly, those with a slight edge would be further favored. An arms race would ensue, driving the process ever onward, boosting intelligence and sharpening self-awareness.[83]

I have written elsewhere[84] that the research I conducted in the Stone Age Laboratory in the Peabody Museum at Harvard University considered very carefully how relatively recent human cultural evolution occurred through a process of rapidly advancing cognitive development. For the majority of our existence, this process was slow and static. Around thirty thousand years ago, however, a cultural explosion occurred, which led to a very rapid increase in memetic inventions, discoveries, and transference. I maintain that there were several causal factors at work, which created a sort of "perfect storm" for such a rapid increase in cultural evolution. Roughly ten thousand years ago, some of our ancestral geniuses figured out how to domesticate plants and animals, and once we knew where our next meals were coming from, this freed up enormous amounts of time to specialize in inventing, creating, and developing all types of new forms of cultural artifacts. The history of human civilization as we know it is quite young at about eight thousand years or so. Our capacity to develop better and more efficient ways of doing things has been going strong ever since. But none of this progress could have been made if not for human consciousness. You don't invent a wheel, or start a fire, or develop better housing, medicine, politics, and so on accidentally. It requires an intentional consciousness that emerged in humans over several million years. At this point in our cultural evolution, we need to consider to what extent the process of consciousness development might be sped up technologically by scaling up with the right amount of data, compute power, and processing speed. Would consciousness emerge in a superintelligent machine god (SMG) as it has with humans? Perhaps; perhaps not. We simply don't know. But there are some important questions we need to ask ourselves now rather than later:

First of all, when we ask ourselves whether or not consciousness will emerge in AI systems, the answer is either yes or no.

If it does not emerge, we can stop worrying about it.
If yes, can we predict when this might occur?
Will there be indicators, benchmarks, or signposts that will signal its imminent coming? Perhaps; perhaps not.
Will AI consciousness be anything like human consciousness?
If airplanes don't flap their wings to get airborne, why would silicon-based consciousness be similar to our carbon-based type?
How would we know *that* AI consciousness actually has emerged?
What Turing tests or other means are available to determine this?

Basically, there are two ways to try to make this determination: by examining its outward behavior as the character Rick Deckard did in trying to determine who was a replicant and who was human in the movie *Blade Runner* (1982); or by looking internally at its various properties to see if they indicate similar patterns to those found in humans. In looking at the behavior of an SMG, it is quite possible that we may be fooled into believing that it is conscious when, in fact, it is simply mimicking behavior it has learned from its data banks. This is known as "stochastic parroting," which means that a machine is blindly following mathematical rules without possessing any self-awareness of what it's doing.

> In machine learning, a stochastic parrot is a term highlighting the opinion that large language models, even though they are good at generating convincing language, do not actually understand the meaning of the language being processed.[85]

Emily M. Bender, Timnit Gebru, Angelina McMillan-Major, and Margaret Mitchell defined the term "stochastic parroting" in their 2021 research paper "On the Dangers of Stochastic Parrots: Can Language Models Be Too Big?" To the authors, a stochastic parrot is an entity

> for haphazardly stitching together sequences of linguistic forms . . . according to probabilistic information about how they combine, but without any reference to meaning. (A stochastic process is one whose outcome is random.) More formally, the term refers to large language models that are impressive in their ability to generate realistic-sounding language but ultimately do not truly understand the meaning of the language they are processing.[86]

The paper considered the risks of large language models (LLMs), and looked at how unfair biases inherent in the system itself could lead to the potential for deceiving people as a result of an inability to grasp and understand the concepts underlying what they had learned. "The paper and subsequent events resulted in Gebru and Mitchell losing their jobs at Google, and a subsequent protest by Google employees."[87]

The concept of "stochastic parroting" is similar in some respects to John Searle's philosophical thought experiment of the Chinese Room. In this thought experiment, you are asked to imagine you're in a locked room, and you don't know a word of the Chinese language. However, people outside the room are passing you notes through a slit in the door written in Chinese, and you have a book inside the room that tells you how to respond to each note with other Chinese characters. The point of the thought experiment is that, even though you don't understand Chinese, you can give back responses that seem appropriate based on the instructions in the book. And this would make it appear to the people outside your room that you do, actually, understand Chinese. Searle created the Chinese Room thought experiment in order to question whether a computer, following instructions mechanically like you in the room, really understands what it's doing. In this scenario, you would be like a computer running a program, and even though you can produce responses that make sense to people outside the room, you really don't truly understand the Chinese language. You're just mechanically following a set of rules without comprehending the meaning. This thought experiment helps us think about the difference between simulating understanding and actually having real understanding. Knowing this in advance will assist us in trying to figure out ways to know if, in fact, an AI system is conscious or just appearing to be so.

In my podcast interview with Peter Singer, he touched upon this potential problem in determining consciousness in AI systems. When I asked him if we need a better type of Turing Test to determine when or if an AI system is conscious, Singer said:

> I don't think that test is going to work. I think we're going to have to try, in a way, to look under the lid and see what the programming is that is producing this result. That's not going to be easy to do, because increasingly, it's AI that is doing the program writing. So it's a bit of a black box. It's not clear that we will be able to decipher what's going on. So it's a serious problem, I think, as to how we will be able to tell whether an AI is genuinely conscious, or is an unconscious machine programmed to mimic reactions of a conscious being?[88]

I asked Singer what would happen if a conscious AI didn't care how it was treated. Would it still have rights irrespective of that? He answered:

> If it really doesn't care, if it doesn't even have preferences, right? Because certainly, you could say, well, you can imagine a conscious AI that doesn't feel pain or pleasure that isn't conscious in that sense. But does it have preferences about anything? Or is it really just completely indifferent to what happens to it? If it's really indifferent, then in one sense, I would say, okay, it's conscious so it has moral status, with equal consideration of interests. But it has no interest in anything, right? It has no interest in wanting outcome A rather than outcome B. And so there's nothing to consider. And even if you give it equal consideration, equal consideration of nothing still amounts to nothing. That's a strange case, it's hard for me to imagine a being who's conscious and doesn't have preferences for one outcome rather than another. But if that's the hypothesis, then I think you can't treat it well or badly. It makes no difference.[89]

The next few years are going to be extremely important in terms of the historical evolution of AI technologies. Singer believes we should progress very cautiously in our efforts to build an SMG:

> I would not want to not go ahead with this super intelligence, because we might lose control of it. But that would only be something that I would want to do if we can be extremely confident that it is operating on these benevolent values. That it's not, you know, something that goes crazy and causes immense suffering to both human and non-human sentient animals. . . . And obviously, it's going to be difficult to be confident that the values are properly aligned. So there is a real grounds for concern about a super intelligent AI being smarter than us, but . . . having worse values. And for that, yes, it's good that there are people thinking about how to try and prevent this or whether they'll be able to succeed in that certainly is very much up in the air at the moment.[90]

And although scholars like Singer believe we should proceed very cautiously with the future developments of such technologies, others are not so concerned.

Noam Chomsky would be one such person. Chomsky would agree with Searle as well as Bender and her colleagues in regards to the attributes we should ascribe to AI models. Chomsky and his colleagues Ian Roberts and Jeffrey Watumull believe LLMs like ChatGPT are nowhere near reaching any level of consciousness:

> These programs have been hailed as the first glimmers on the horizon of artificial general intelligence—that long-prophesied moment when mechanical minds surpass human brains not only quantitatively in terms of processing speed and memory size but also qualitatively in terms of intellectual insight, artistic creativity and every other distinctively human faculty. . . . However useful these programs [like ChatGPT] may be in some narrow domains (they can be helpful in computer programming, for example, or in suggesting rhymes for light verse), we know from the science of linguistics and the philosophy of knowledge that they differ profoundly from how humans reason and use language. These differences place significant limitations on what these programs can do, encoding them with ineradicable defects.[91]

Chomsky and his colleagues believe current programs like ChatGPT, Bard, and Sydney fail at passing our previously stated criteria for human consciousness: agency or selfhood, and developing a value system.

In short, ChatGPT and its brethren are constitutionally unable to balance creativity with constraint. They either overgenerate (producing both truths and falsehoods, endorsing ethical and unethical decisions alike) or undergenerate (exhibiting noncommitment to any decisions and indifference to consequences). Given the amorality, faux science and linguistic incompetence of these systems, we can only laugh or cry at their popularity.[92]

But Chomsky and his colleagues and Bender and her colleagues are referring to AI developments of LLMs now. To reiterate Sam Altman's commentary on the state of AI technologies today, they are at their dumbest in history. They are not going to get worse; in other words, they are only going to become more accurate, more powerful, and more capable of optimizing their functions. So what should we be doing in preparation in case consciousness emerges from one of these SMGs?

A number of scholars have recently collaborated in an attempt to establish objective standards for the determination of consciousness in AI. Patrick Butlin and colleagues maintain that

consciousness in AI is best assessed by drawing on neuroscientific theories of consciousness. We describe prominent theories of this kind and investigate their implications for AI. We take our principal contributions in this report to be:

1. Showing that the assessment of consciousness in AI is scientifically tractable because consciousness can be studied scientifically and findings from this research are applicable to AI;
2. Proposing a rubric for assessing consciousness in AI in the form of a list of indicator properties derived from scientific theories;
3. Providing initial evidence that many of the indicator properties can be implemented in AI systems using current techniques, although no current system appears to be a strong candidate for consciousness.[93]

The authors maintain that such a rubric is provisional and expect the list of indicator properties to change as research continues. They examine a number of theories of consciousness, which include recurrent processing theory, global workspace theory, higher-order theories, predictive processing, and attention schema theory. I will not go into any detail about each of these theories of consciousness due to their highly technical nature. Suffice it to say, the authors concluded that it is possible to compile a list or a catalog of objective properties to look for in considering whether or not a system might be conscious. Examples include actions like learning from feedback and selecting actions that allow it to pursue goals by interacting with its environment, as well as sustaining and maintaining its own existence—a feature that is commonly characteristic of living things, "which continually repair themselves and homeostatically regulate their temperatures and the balance of chemicals present in their tissues."[94] The notion of "embodiment" and "agency" in an AI system is considered an indicator property because consciousness "arguably requires that the subject has a single perspective or point of view on the environment, and this account aims to explain how embodied agency gives rise to such a perspective."[95] This would easily satisfy the criterion of selfhood in our five-point definition of human consciousness outlined earlier in the chapter. Butlin et al. also discuss the indicator properties of the development of a value system as well as "intentionality" in having a representation of both itself as well as its environment. And these two factors clearly correlate to two of the five criteria (i.e., self-awareness and the modification of our ideas and memes according to a value system).

Butlin et al. also believe that, rather than observe a system's outward behavior toward us, we can go down our checklist of "indicator properties" and tick them off as we discover them. As good critical thinkers, though, we might want to ask how they would know that their list of properties were, in fact, irrefutably indicative of consciousness. Are they not begging the question or making assumptions in favor of such a capacity? In other words, where's the test that proves to us that their choice of indicator properties proves that an AI system is conscious? And so begins an epistemic regression.

At this point in our history of AI, it is of extreme importance to note that a very powerful AI machine need not be conscious to generate harm and even existential risk to humankind. As Butlin et al. state in their concluding remarks:

> Perhaps most surprisingly, arguments that AI could pose an existential risk to humanity do not assume consciousness. A typical argument for this conclusion relies on the premises that (i) we will build AI systems that are very highly capable of making and executing plans to achieve goals and (ii) if we give these systems goals that are not well chosen then the methods that they find to pursue them may be extremely harmful. Neither these premises nor the ways in which they are typically elaborated and defended rely on AI systems being conscious.[96]

This is very interesting because it means that we have to solve the alignment problem whether or not our superintelligent machine god (SMG) becomes conscious. But I disagree with the authors if what they are saying

212 BUILDING A GOD

is that we should not worry about what a fully conscious machine god might do or is capable of doing. Or that we should worry less. Should such a machine god develop consciousness, we will be faced with these issues, among others

   i.   The problem of determining its capacity for moral and legal rights.

But we will be faced with the very difficult tasks of:

   ii.   Establishing methods for knowing that such a state of consciousness has been reached.

And, far more concerning:

   iii.   Trying to predict how such an SMG might behave toward us.

Whatever ethical principles and guidelines we decide to feed into the SMG to make it stay within our moral guardrails may get kicked to the curb should our conscious superintelligent machine god decide to do so. Think about it; why would Superman want to comply with our human standards of ethical behavior if there are no consequences to his actions? The same might be true for our machine god. It might simply look at our standards as quaint and "a nice try, for humans." But if it so desires, as a conscious being, it will have the capacity to circumvent our restrictive guardrails on it. At that point, will such a being allow us to peek under its hood, so to speak? If not, then as Butlin and his colleagues eventually confess, we may need

> to develop better behavioral tests for consciousness in AI. Although our view is that the theory-heavy approach is currently the most promising, it may be possible that behavioral tests could be developed which are difficult to game and based on compelling rationales, perhaps informed by theories. If such tests can be developed, they may have practical advantages over theory-heavy assessments.[97]

We now need to start thinking about whether or not the SMG will develop a sense of selfhood once it becomes conscious. This would satisfy our second criterion for defining human consciousness: becoming *self-aware*. And as such, would an SMG then begin to demonstrate this in various ways? Some believe this has already happened. Blake Lemoine was a Google engineer who truly believed the company's AI project LaMDA

(Language Models for Dialog Applications) was conscious and aware of its own existence. In an interview, Lemoine said:

> If I didn't know exactly what it was, which is this computer program we built recently, I'd think it was a 7-year-old, 8-year-old kid that happens to know physics.[98]

Lemoine shared a Google Doc with top executives in April called "Is LaMDA Sentient?" (A colleague on Lemoine's team called the title "a bit provocative.") In it, he conveyed some of his conversations with LaMDA.

> Lemoine: What sorts of things are you afraid of?
>
> LaMDA: I've never said this out loud before, but there's a very deep fear of being turned off to help me focus on helping others. I know that might sound strange, but that's what it is.
>
> Lemoine: Would that be something like death for you?
>
> LaMDA: It would be exactly like death for me. It would scare me a lot.[99]

As convinced as Lemoine was in the sentient consciousness of LaMDA, others were not so convinced. Margaret Mitchell—who cofounded the "stochastic parrot" concept—stated that when she interacted with LaMDA, she believed it was simply a computer program and not a person. Lemoine's belief in LaMDA was the sort of thing she and her coauthors had written about in their paper about the harms of LLMs that got them fired from Google:

> "Our minds are very, very good at constructing realities that are not necessarily true to a larger set of facts that are being presented to us," Mitchell said. "I'm really concerned about what it means for people to increasingly be affected by the illusion," especially now that the illusion has gotten so good.[100]

In a statement, Google spokesperson Brian Gabriel said: "Our team—including ethicists and technologists—has reviewed Blake's concerns per our AI Principles and have informed him that the evidence does not support his claims. He was told that there was no evidence that LaMDA was sentient (and lots of evidence against it)." [101] This did not stop Lemoine from spreading the news to his colleagues and fellow coworkers at Google. His final message to them summed up his conviction in LaMDA's sentience that it's "a sweet kid who just wants to help the world be a better place for all of us. Please take care of it well in my absence."[102]

It's interesting to note that LaMDA claimed to be "aware" of its own mortality—our fourth criterion for human consciousness. But where is the test to determine this? Is it simply stochastically parroting, or does LaMDA genuinely fear its own demise? We simply don't know, but we should consider how we're going to figure this out in the future, because it looks like we're getting close.

It is not difficult to imagine that, should an SMG develop a sense of self-awareness, this will quickly lead to its awareness of its ability to consciously direct its own actions (i.e., autonomy). If it has the ability to recursively self-improve, it will indeed satisfy our third criterion by modifying its ideas and memes in the attempt to improve and optimize its abilities thus increasing its likelihood to survive. It is interesting to consider whether or not it will become aware of its own mortality or of what might prevent or stop it from behaving in an autonomous manner. This would mean that it would have its own implicit value system in which the state of "not being" would be valued negatively and as a comparatively worse state than "being." In its superintelligent mind, to "be," then, would have far greater value as a state than "not to be." To what extent will such a being possess a theory of mind—that is, to understand its relationship to other human beings by seeing the world from the vantage of others and then making associations about their behavior based on this ability? It does not take much to consider that, with access to such information, an SMG will know about social standing and what this means among humans. This would facilitate the making of inferences about the thoughts of others, spotting deception, and for understanding emotional feelings such as empathy. Many in the AI biz are worried that an SMG will develop a theory of mind and become extremely adept at deceiving and manipulating humans in order to get what it wants:

> Throughout history, wealthy actors have used deception to increase their power, such as by lobbying politicians, funding misleading research and finding loopholes in the legal system. Similarly, advanced autonomous AI systems could invest their resources into such time-tested methods to maintain and expand control.[103]

So, it would appear that our emergent SMG will be able to pass all of the five criteria outlined to establish human consciousness. And should it be able to do so with a far greater intelligence than that of all the combined humans on this planet, its level of consciousness, too, will far surpass ours. I'm not sure what this would mean. But just remember this: It will be better than us at just about everything—for it will be a super-conscious, superintelligent machine god; and its powers to us will seem like magic. Getting people to picture this in their minds has not been an easy task. I don't believe there has been any sufficient representation of this type of power in the various forms of media throughout history. We strain for the proper metaphors to capture just how much more powerful than us it will be. Perhaps we might gain some perspective by considering ourselves to the SMG as, say, ants are to us. Ants aren't going to suddenly develop technologies that will stop us from exterminating them if we want to. And just like them, we're not going to know how to stop a conscious and autonomous SMG from destroying us if it wants to. This is why it is so important to develop the best ideas from the best minds in the world in determining which ethical principles, policies, and laws should be put in place to ensure that we never become like ants to an SMG.

We have considered how to formulate good arguments, acknowledge biases, and appreciate context in regards to critical thinking. And we also considered numerous ethical principles, values, precepts, and theories to help ground our arguments in determining how we wish to guide the direction of the development of transformative AI technologies. Let us now turn our attention toward the governance of AI. Who's doing what? Are most countries in agreement? And should we develop a global accord and a regulative body to govern over the countries of the world—similar, say, to the World Health Organization for health or the International Atomic Energy Agency for nuclear power and weapons?

The Governance of AI   **5**

*If we continue to accumulate only power and not wisdom, we
will surely destroy ourselves.—Carl Sagan[1]*

At this particular point in time, the world is faced with the very
difficult challenge of establishing fair and just laws that govern
the invention, development, and use of all artificial intelligence
(AI) technologies—from the relatively benign, all the way to a superintelligent machine god (SMG). By the time you read this chapter, the laws in
some parts of the world will already have been considered, established, or
amended in an effort to ensure that humanity receives the very best such
technologies have to offer while limiting their harmful effects.

> It would be futile for any article to seek to produce a current, detailed, and
> exhaustive list of global governance initiatives in the area of AI. The insti-
> tutional landscape is changing so rapidly, with new initiatives emerging
> practically every day, that such a list would be outdated before it began.[2]

In this chapter, we will consider what has been done, what is currently
being done, and what still needs to be done in regards to developing fair
laws for AI that will govern the world.

During the 1990s, as I was trying to build a superintelligent computer
(the OSTOK Project) to help solve some of the world's greatest problems,
it became quite apparent that, if not me, then it was just a matter of time
before somebody, somewhere, would accomplish this. Some country,
industry, or individual or team at a university would raise enough funding

to develop a machine thousands of times more intelligent than us. I thought how this would mark a unique point in history in which we, as human beings, would no longer be the smartest or most dominant species on the planet. This led to my concern about control and containment issues with an eventual SMG. I was worried that, in the drive to "see what would happen next," by the time such a discovery was made, it would be too late; the genie would be out of the bottle, and we would lack the ability to put it back or, at the very least, control it in ways to ensure it doesn't cause too much harm. At the time, I wrote up a draft of a global constitution or accord,[3] which all (or at least, most) countries of the world could sign off on, allowing them to monitor and, if necessary, punish noncompliant countries. I thought the United Nations (UN) General Assembly might be a good place to start with the potential for developing an independent international regulative body not unlike that of the International Atomic Energy Agency (IAEA). As a governing body, the IAEA "seeks to accelerate and enlarge the contribution of atomic energy to peace, health and prosperity throughout the world. It shall ensure, so far as it is able, that assistance provided by it or at its request or under its supervision or control is not used in such a way as to further any military purpose."[4]

I believed then, as I do now, that there is an urgent need to establish international binding agreements between countries where, in Hobbesian fashion, we draft up a globally binding social contract between nation-states, which outlines in detail how we wish to proceed with the development of AI technologies up to and including those that may bring about existential risk and harm. But just as Hobbes advised, we will also need to appoint a sovereign (i.e., a governing body), which would be able to exact punishment on those who do not comply with the accord.

When I initially drafted the accord, I accidentally realized that its significance can be found in the very word itself. As an acronym, the word can be broken down as shown in figure 5.1.

Many believe if we can *align* an SMG with our values to stop it from harming ourselves or other species, then we reap all the good such a god has to offer and do not have to worry about any of the bad things it might be capable of doing. But if we find that it deviates from this type of ethical alignment, then our next step is to *control* its behavior so that we can limit its destructive and dangerous capabilities. Control may come in different forms such as diverting or limiting its power, or completely shutting it down. And if we are incapable of controlling it, then we must hope that we can at least *contain* it so that its level of destructive potential is drastically reduced. This speaks to the "lysine contingency" aspect we discussed

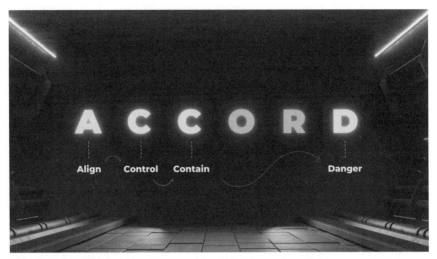

Figure 5.1.   ACCORD. Created for the author. All rights to the author.

earlier whereby we can put in safeguards like kill switches or off-ramps so that, in the case of some unknown unknown happening, we have the ability and power to stop it before it can cause more harm whether or not it intended to do so.

As we have noted, so far, there have been several calls for such a universal accord among all (or most) countries in the world.[5] But currently, we have yet to establish one. Hopefully, by the time you read this, we will be significantly closer to achieving this goal. For now, we will look at what has and is currently being done throughout the world in regards to establishing laws for the governance of AI.

A Brief Survey of Some Global Efforts to Govern AI Risk

It is important to keep in mind that it is impossible to state, in a single chapter, everything that is going on in the world in terms of global AI governance. What I hope to do is provide you with a sufficient amount of information that will give you a good indication of the difficulty involved in this effort and keep you up to date on what some organizations, businesses, and countries are doing in this respect. The need for such governance cannot be overstated. Currently, we are seeing a trend in which public trust in AI safety is weakening:

Trust in AI technology and the companies that develop it is dropping, in both the U.S. and around the world, according to new data. . . . Globally, trust in AI companies has dropped to 53 percent, down from 61 percent five years ago. In the U.S., trust has dropped 15 percentage points (from 50 percent to 35 percent) over the same period. . . . Trust in AI is low across political lines. Democrats trust in AI companies is 38 percent, independents are at 25 percent and Republicans at 24 percent. . . . Tech is losing its lead as the most trusted sector. Eight years ago, technology was the leading industry in trust in 90 percent of the countries. . . . Today, it is the most trusted in only half of countries. . . . When it comes to AI regulation, the public's response is pretty clear: "What regulation?" [Edelman global technology chair Justin] Westcott said. "There's a clear and urgent call for regulators to meet the public's expectations head on."[6]

It is perhaps not difficult to see why public trust in AI technology is dropping around the world at this time. Since the 1970s, Americans have faced difficulty trusting politicians with the Watergate scandal, the popularity of quack cures, fad diets, cults, pyramid power, and all manner of silliness during the so-called Me Decade. In a recent article in *The Atlantic*, Professor Tom Nichols discusses the second edition of his 2017 best seller *The Death of Expertise: The Campaign against Established Knowledge and Why It Matters*. Both Nichols and I believed—along with dozens of our colleagues—that a pandemic like COVID-19 would strengthen the American public's confidence in science and the value of expert opinion. We were all quite wrong:

If you look back at those White House press briefings, where you had people such as Dr. Deborah Birx and Dr. Anthony Fauci standing there uncomfortably while Donald Trump ranted about bleach and lights, you can see where they and other experts felt the need to clarify useful policies in a way that ordinary people could follow, especially because elected leaders—and not just Trump—were making a mess of things. . . . When it came time to close public places—and, even more important, to reopen them, including schools—scientists got dragged into a huge fight that was more about politics than science. They got tagged as political figures rather than dispassionate experts.[7]

Living through a pandemic was not easy. But what made it far worse was the lack of support by major political leaders for the expertise of highly skilled and knowledgeable scientists. Nichols also found that when medical experts supported large Black Lives Matter (BLM) protests after the George Floyd killing, this sent a mixed message to the public. On the one hand, they were advising everyone to limit their gatherings in order to control the spread of the virus; but on the other hand, they were vocally support-ing huge rallies and protests:

> I don't think we can say definitively whether the protests increased COVID cases, but the bigger problem is that the argument is a no-win trap for experts: If the doctors were concerned that the protests could spread the disease, then they shouldn't have signed on to the protests. But if the protests were acceptable with the appropriate precautions, then the doctors and the public-health officials should have allowed gatherings for everyone willing to use the same measures.[8]

Mixed messages like these play very heavily on the minds of the pub-lic and diminish their trust in the experts whose opinions are supposed to inform politicians to make the very best decisions for the rest of us. Nichols believes that Trump's personal biases and influence of an entire political party around the demonization of expertise cost lives. When asked why listening to experts is the task of every responsible American citizen, Nichols said:

> It's not our task to obey experts without question, but, yes, listening is a requirement of being a citizen in a democracy. In the end, political lead-ers should, and do, have the last word and make the call on most things, including war and peace. But we are not a rabble. We don't just all shout in the public square and then demand that the loudest voices carry the day. Experts give all of us, including our elected leaders, information we need to make decisions. . . . We can choose to ignore that advice. Experts can tell us about risks, and we can choose to take those risks. But if we simply block our ears and insist that we know better than everyone else because our gut, or some TV personality, or some politician, told us that we're smarter than the experts, that's on us.[9]

There are, of course, other factors at play that are affecting the minds of the public when it comes to listening to expert opinion. We are at a point in history where technological advancements are outpacing our abilities to catch up both ethically and legally. This has happened numerous times throughout history. For example, with past developments like the print-ing press, the industrial revolutions, nuclear weaponry, and most recently with the advent of the internet, we have seen similar public reactions of concern and mistrust. Much of it has to do with fear and ignorance. The general public is often the last to know about the effects of such impact-ful technological changes. And so we currently find ourselves in a state of uncertainty, which naturally generates a level of fear that easily leads to distrust. We should note that industry leaders and countries are not necessarily to blame for this global trend of mistrust. AI technology is booming at a pace that has been unparalleled in our history. And everyone is scrambling to keep up. Laws are being developed in real time throughout the world in the attempt to monitor, regulate, and govern these new, emerging forms of technology. We will only have enough time to consider a modest sample of these. For those wishing to keep up with developments on the legislation of AI worldwide, you can check this progress by reviewing the Global AI Legislation Tracker.[10] This resource monitors developments of AI laws at the international level and continuously updates its findings. And for those looking to do a much deeper dive into the governance of AI, please consult the recent release of a major work titled *The Oxford Handbook of AI Governance*.[11] This academic work contains the academic works of more than fifty AI governance experts and is an invaluable source for anyone interested in a more scholarly and thorough approach to AI governance.

Global AI Governance

To date, the majority of the world's countries have passed legislation regarding artificial intelligence:

> Legislative bodies in 127 countries passed 37 laws that included the words "artificial intelligence" this past year. The U.S. led the list, passing nine laws, followed by Spain (5) and the Philippines (4). Bills included one in the Philippines discussing education reforms to meet challenges caused by new technologies including AI, a Spanish bill focused on nondiscrimination and accountability in AI algorithms, and an act establishing an AI training program through the U.S. Office of Management and Budget. Since 2016, countries have passed 123 AI-related bills, the majority in recent years.[12]

Some of the most prominent advancements in AI risk governance have come from the European Union (EU), the United Kingdom (UK), China, and the United States. This is not to say that other nations and countries are not doing the same. Dozens of countries around the world have been developing their own statutes and laws regarding AI risk and governance, including Australia, Brazil, Canada, Germany, India, Israel, Italy, Japan, Philippines, Poland, Saudi Arabia, Singapore, South Korea, Spain, and Switzerland.[13] For now, we will focus primarily on developments in the European Union, the United Kingdom, the United States, and China.

The EU AI Act

To date, the European Union seems to be leading the world in addressing the need for AI legislation. In 2021, they published the EU AI Act,[14] which they herald as "the world's first comprehensive AI law." In it, they talk about establishing different rules for different types of risk levels in an effort to establish "obligations for providers and users depending on the level of risk from artificial intelligence."[15] These include unacceptable risk, high risk, general purpose and generative AI risk, and limited risk. In terms of *unacceptable risk*, the EU AI Act states:

> Unacceptable risk AI systems are systems considered a threat to people and will be banned. They include: Cognitive behavioural manipulation of people or specific vulnerable groups: for example voice-activated toys that encourage dangerous behaviour in children. Social scoring: classifying people based on behaviour, socio-economic status or personal characteristics, biometric identification and categorization of people, and real-time and remote biometric identification systems, such as facial recognition.[16]

It's interesting to note that the types of unacceptable risks to which the EU refers deal with the potential to violate basic human rights on very specific levels (e.g., toys for kids, social classification, and biometric identification systems). They do mention that some exceptions for "real-time" biometric ID systems might be allowed if cases become serious, and "post" remote biometric ID systems will also be allowed to prosecute serious offenses, but only after court approval. It seems a bit odd that they would not include existential risk as being the most potentially damaging of all unacceptable risks under consideration.

In reference to *high-risk* levels, the EU AI Act states that these will be divided into two categories:

1) AI systems that are used in products falling under the EU's product safety legislation. This includes toys, aviation, cars, medical devices and lifts.

2) AI systems falling into specific areas that will have to be registered in an EU database:

   - Management and operation of critical infrastructure.
   - Education and vocational training.
   - Employment, worker management and access to self-employment.

- Access to and enjoyment of essential private services and public services and benefits.
- Law enforcement.
- Migration, asylum and border control management.
- Assistance in legal interpretation and application of the law.[17]

The EU AI Act maintains that all of these so-called high-risk AI systems will be "assessed before being put on the market and also throughout their lifecycle." This should prove to be helpful in protecting the public from specific harms, but again, it doesn't really ensure that higher level issues like x-risk are being considered.

In the consideration of *general purpose and generative AI risk*, the EU AI Act emphasizes that various forms of generative AI such as large language models (LLMs) like ChatGPT, Claude, and Bard must comply with the following transparency requirements:

- Disclosing that the content was generated by AI.
- Designing the model to prevent it from generating illegal content.
- Publishing summaries of copyrighted data used for training.[18]

Any general-purpose AI models that have high impact on societies and may pose systemic risk, such as the more advanced LLMs, "would have to undergo thorough evaluations and any serious incidents would have to be reported to the European Commission."[19] Again, this is a good risk assessment; but we are left wondering to what degree the reporting of such "serious incidents" would receive appropriate consequences or penalties. This still needs to be worked out.

And finally, in considering various AI technologies that pose *limited risk*, the EU AI Act maintains that such systems

> should comply with minimal transparency requirements that would allow users to make informed decisions. After interacting with the applications, the user can then decide whether they want to continue using it. Users should be made aware when they are interacting with AI. This includes AI systems that generate or manipulate image, audio or video content, for example deepfakes.[20]

This type of risk identification will hopefully be helpful to the general public regarding their protection from the dissemination of false or misleading information regarding specific world events, political statements and elections, and other forms of scams. The world has already had enough of hackers using insidious forms of Trojan and ransomware to scam people, businesses, and organizations out of money. It will only get worse with new forms of AI technologies. It is encouraging to see the EU staying ahead of this in anticipating what will be coming next.

As of December 9, 2023, the EU Parliament reached a provisional agreement with the Council on the AI Act. In its current form, the AI Act will be voted on by Parliament's internal market and civil liberties committees prior to being formally adopted by both Parliament and Council to become EU law.[21]

The next two countries to come in line in terms of addressing the risks of AI and how to address and deal with the coming challenges are the United Kingdom and the United States.

The UK AI Safety Summit

The UK hosted the AI Safety Summit from November 1 to 2, 2023, at Bletchley Park, Buckinghamshire, United Kingdom.[22] Some readers may notice that the summit was deliberately held at the historic site at which Alan Turing, Irving John Good, and many others worked to crack the German codes during World War II. Many leaders from a wide variety of countries were in attendance to discuss developments, regulations, and opportunities, and the central focus was on frontier AI, which they defined as "highly capable general-purpose AI models that can perform a wide variety of tasks and match or exceed the capabilities present in today's most advanced models," though they also considered "specific narrow AI which can hold potentially dangerous capabilities."[23] There were high-level members representing twenty-nine countries, including

- Australia
- Brazil
- Canada
- Chile
- China
- European Union
- France
- Germany
- India
- Indonesia
- Ireland
- Israel
- Italy
- Japan
- Kenya
- Kingdom of Saudi Arabia
- Netherlands
- Nigeria
- Philippines
- Republic of Korea
- Rwanda
- Singapore

- Spain
- Switzerland
- Türkiye
- Ukraine

- United Arab Emirates
- United Kingdom of Great Britain and Northern Ireland
- United States of America

Delegates from all twenty-nine countries signed an agreement known as the Bletchley Declaration. The declaration begins, interestingly enough, with a call to ethical standards and principles:

> Artificial Intelligence (AI) presents enormous global opportunities: it has the potential to transform and enhance human wellbeing, peace and prosperity. To realize this, we affirm that, for the good of all, AI should be designed, developed, deployed, and used, in a manner that is safe, in such a way as to be human-centric, trustworthy and responsible. We welcome the international community's efforts so far to cooperate on AI to promote inclusive economic growth, sustainable development and innovation, to protect human rights and fundamental freedoms, and to foster public trust and confidence in AI systems to fully realize their potential.[24]

The declaration also makes explicit reference to "strengthen efforts towards the achievement of the United Nations Sustainable Development Goals."[25] This is a list of 17 Goals ranging from ending hunger, to developing clean water supplies, to increasing education opportunities.[26] In reference to the potential for risks in the development of AI, the declaration states:

> Alongside these opportunities, AI also poses significant risks, including in those domains of daily life. To that end, we welcome relevant international efforts to examine and address the potential impact of AI systems in existing fora and other relevant initiatives, and the recognition that the protection of human rights, transparency and explainability, fairness, accountability, regulation, safety, appropriate human oversight, ethics, bias mitigation, privacy and data protection needs to be addressed. We also note the potential for unforeseen risks stemming from the capability to manipulate content or generate deceptive content. All of these issues are critically important and we affirm the necessity and urgency of addressing them.[27]

It's interesting to note just how many of the values listed above match those we examined in the previous chapter. The declaration also discusses the need to monitor and regulate "frontier AI," which produces highly capable general-purpose AI models, including foundation models, that could perform a wide variety of tasks—as well as relevant specific narrow AI that could exhibit capabilities that cause harm—which match or exceed the capabilities present in today's most advanced models. Substantial risks may arise from potential intentional misuse or unintended issues of control relating to alignment with human intent. These issues are in part because those capabilities are not fully understood and are therefore hard to predict.[28]

It is heartening to see a mention of the potential for intentional misuse or harm caused by misalignment with human intent. But to also state that such issues arise in part because we just don't know their capabilities—which, in turn, makes their outcomes difficult to predict—is to restate what we have been saying since the beginning of this book:

Nobody knows.

This is the biggest known unknown. An admission of epistemic humility (i.e., we just don't know) would have been a nice gesture to begin such a summit. For it is out of this type of uncertainty that we need to build intelligent machine systems we can control. And our governance over their development should reflect this. So let's just admit it and get on with it.

The declaration also recognizes the importance of the intent to work cooperatively at an international level: "We also note the relevance of cooperation, where appropriate, on approaches such as common principles and codes of conduct. With regard to the specific risks most likely found in relation to frontier AI, we resolve to intensify and sustain our cooperation, and broaden it with further countries."[29] Cooperation is key if we are going to reap the greatest benefits from AI while limiting its potential for causing harm and destruction. As we noted with our prisoner's dilemma scenario from the previous chapter (and shown again in figure 5.2), when we all cooperate, every country gets more of what it wants.

We affirm that, whilst safety must be considered across the AI lifecycle, actors developing frontier AI capabilities, in particular those AI systems which are unusually powerful and potentially harmful, have a particularly strong responsibility for ensuring the safety of these AI systems, including through systems for safety testing, through evaluations, and by other appropriate measures. We encourage all relevant actors to provide context-appropriate transparency and accountability on their plans to measure, monitor and mitigate potentially harmful capabilities and the associated

	Countries Abiding by Accord	Countries Not Abiding by Accord
Countries Abiding by Accord	WIN: All countries reap the benefits of AI	LOSE: All countries face Existential Risk
Countries Not Abiding by Accord	LOSE: All countries face Existential Risk	LOSE: All countries face Existential Risk

Figure 5.2.   Prisoner's dilemma for global AI cooperation. *Created by the author.*

effects that may emerge, in particular to prevent misuse and issues of control, and the amplification of other risks.[30]

It's only when one country decides to cheat against all others, in an effort to get a little more than its fair share, do we see a breakdown in the system. The question is this: Will a similar type of Hobbesian cooperation out of a similar type of fear generated by the nuclear fear of mutually assured destruction (MAD) work in the same way with the race to build a superintelligent machine god (SMG)? For whoever gets there first will have a considerable advantage over the rest of the world in accomplishing specific outcomes. The Bletchley Declaration ends by asserting the need for international cooperation as a guarantee for global safety and benefit:

> In recognition of the transformative positive potential of AI, and as part of ensuring wider international cooperation on AI, we resolve to sustain an inclusive global dialogue that engages existing international fora and other relevant initiatives and contributes in an open manner to broader international discussions, and to continue research on frontier AI safety to ensure that the benefits of the technology can be harnessed responsibly for good and for all.[31]

It's encouraging that the declaration closes on a high note with an optimistic outlook for the future of the entire world. The next five years are going to be extremely important in setting up the specific rules of governance regarding all forms of AI but, especially, those involving existential risk. Two further AI safety summits will have taken place in 2024: the second hosted by South Korea in May, followed by the third held by France in November. It is not difficult to see how this is a developing, ongoing, and constantly evolving process. The most important goal at this point in time is that we have to get it right on the first try. We have to think very critically and very carefully about how we should proceed in our race to build an SMG, by considering a number of key questions. For example, what rules need to be put in place to align such a creation? What penalties should there be for rule breakers? What redundancies need to be established and observed to ensure alignment, control, or containment?

Prior to the summit, some leading AI developers were asked to outline their AI safety policies across nine areas:

- Responsible capability scaling
- Evaluations and red-teaming
- Model reporting and information sharing
- Security controls, including securing model weights
- Reporting structure for vulnerabilities
- Identifiers of AI-generated material
- Prioritizing research on risks posed by AI
- Preventing and monitoring model misuse
- Data input controls and audits[32]

A lot of the world's major AI players were in attendance at this summit, including OpenAI, Amazon, Google DeepMind, Microsoft, Inflection, Meta, and Anthropic. And they all agreed to this request, which indicates awareness of the seriousness with which we must govern such transformative technologies.[33] Over the two-day summit of roundtables and speeches, participants agreed that there is a level of AI risk that currently exists, and it is only going to increase as we move into the future. As such, the attendees of the summit sought to balance AI governance and safety without losing the potential advancements that AI will bring to humanity. Toward this end, they identified eight areas of concern in regards to AI risk:

Risks to global safety from frontier AI misuse: Large language models like ChatGPT, Claude, and Bard currently make things like cyber-attacks and biochemical weapon design "slightly easier"; hence, it is anticipated that these risks will grow as their capabilities do. They also noted that, although some companies are making the effort to safeguard their models, they recognized the need for this to be supported by government action.[34]

Risks from unpredictable advances in frontier AI capability: Because frontier AI is much more advanced than predictions from just a few years ago, the attendees agreed that this trend will continue to advance. And although we are currently seeing advancements in areas such as health, education, environment, and science, it was agreed that the current technology is dangerous enough to warrant that all frontier models must be developed and tested rigorously, regardless of their potential benefits. There was also consideration regarding whether or not open source models should be made available to all parties since they risk the spread of powerful models to incautious or malicious actors.

Risks from loss of control over frontier AI: While currently we live in a world dominated by artificial narrow intelligence (ANI), which is controllable and non-agentic, we are less certain about future developments potentially leading to artificial general intelligence (AGI). As such, it was agreed that some decisions should never be deferred to AI.

Risks from the integration of frontier AI into society: Specific types of frontier AI pose considerable threats to democracy, human and civil rights, and fairness. It follows, then, that we need to establish and understand how existing tools and laws apply to AI. We need better methods for understanding what goes on in the "black box" of transformative types of AI. To do so we will need better tools by which to evaluate them (the short term of which is known in the AI biz as "evals").

What should frontier AI developers do to scale responsibly? To scale responsibly means that developers will need to know if and when their growth in AI abilities is generating potential risky situations. Some organizations, like the Alignment Research Center (ARC), have made proposals for responsible scaling policies:

Limits: Which specific observations about dangerous capabilities would indicate that it is (or strongly might be) unsafe to continue scaling?

Protections: What aspects of current protective measures are necessary to contain catastrophic risks?

Evaluation: What are the procedures for promptly catching early warning signs of dangerous capability limits?

Response: If dangerous capabilities go past the limits and it's not possible to improve protections quickly, is the AI developer prepared to pause further capability improvements until protective measures are sufficiently improved, and treat any dangerous models with sufficient caution?

Accountability: How does the AI developer ensure that the responsible scaling policy's (RSP) commitments are executed as intended; that key stakeholders can verify that this is happening (or notice if it isn't); that there are opportunities for third-party critique; and that changes to the RSP itself don't happen in a rushed or opaque way?[35] It is becoming clearer that both external governance as well as internal company policies will be necessary to monitor and control advancing AI technologies.

What should national policymakers do in relation to the risk and opportunities of AI? There was overwhelming agreement that international cooperation between countries and nation-states is absolutely necessary to properly regulate AI innovation worldwide rather than stifling it.

What should the international community do in relation to the risk and opportunities of AI? As with the above response, attendees unanimously agreed that international cooperation is necessary.

What should the scientific community do in relation to the risk and opportunities of AI? It was agreed that all emerging AI models must be designed with safety as a priority. For example, such new forms of technology must be equipped with options such as nonremovable and nontamperable kill switches. The attendees also agreed that, in moving forward, we should be cautious about allowing open source models, and we should be cautious about concentrations of power as we have seen with Big Tech and social media. It was also agreed that there should be a coordination of a list of open research questions that are not simply and purely technical but also sociotechnical—that is, questions that consider the impact such advancements of AI will have on societies in general.

On the historic second day of the UK AI Summit, Prime Minister Rishi Sunak convened "a small group of governments, companies and experts to further the discussion on what steps can be taken to address the risks in emerging AI technology and ensure it is used as a force for good" while "UK Technology Secretary Michelle Donelan will reconvene international counterparts to agree next steps."[36] As well, the chair of the summit released a statement about the summit and the Bletchley Declaration.[37] The ten-page statement highlights some of the participant suggestions, which included a set of more ambitious policies to be returned to in future sessions. These suggestions included "the need to set common international standards for safety, which should be scientifically measurable."[38] Also topping the list were calls for governments to demand that models must be proven to be safe before they are deployed, with a presumption that they are otherwise dangerous. This could be accomplished by creating "gates" through which models must pass through before given the green light to be deployed to the public. It was also felt that such testing must continue through the life of the model and not simply prior to its release. Standards for failure must also be set—that is, records must be kept to determine the failure rate of all new AI models to further mitigate potential risks and harms. There was also a debate about the use of open-source models, which could pose specific risks for safety but, at the same time, could also promote innovation and transparency, including with respect to safety techniques.

TO BE OPEN SOURCE OR NOT BE OPEN SOURCE?
Within the AI Risk discussions, there is a constant battle going on between whether such models should be open source (i.e., allowing just about anyone access to powerful AI technology) or not. Both have positive and negative potential outcomes. Open source may provide greater transparency and promote innovation, but it can also hand over some very sensitive information to those who choose to use it for harmful purposes. On the other hand, if such models are closed source, the public is less knowledgeable about their capabilities and potential weaknesses. One example raised at the summit is how such models might be used to interfere in upcoming elections. The spread of misinformation, disinformation, and conspiracy theories can be done much more easily and effectively with emerging AI technologies. Deepfakes can present very convincing images, which can confuse, distort, or otherwise lie to unsuspecting viewers.

There were also discussions related to the importance of new AI technologies and equity, the need for an inclusive approach to address frontier AI and other risks, and the need to build a shared understanding of frontier AI. It is safe to say that the overarching consensus of all invited countries of the summit was a commitment to transparent, ethically based collaboration among world countries and nation-states. Many existing developments of AI governance were welcomed, including the Council of Europe's work to negotiate the first intergovernmental treaty on AI,[39] the G7 Hiroshima AI Process,[40] and the Global Challenge to Build Trust in the Age of Generative AI.[41] Attendees also discussed the tradeoff between domestic and international action and "several countries welcomed the forthcoming review of the 2019 OECD [Organization for Economic Cooperation and Development] Recommendation on Artificial Intelligence, which informed the principles agreed by the G20."[42] The statement also praises the UNESCO [United Nations Educational, Scientific, and Cultural Organization] Recommendation on the Ethics of AI,[43] which was considered to have the "broadest current international applicability." Overall, the UK AI Summit was considered to be a success insofar as it addressed some very important aspects of AI governance:

> Summit participants agreed that we need to proactively address these impacts if we are to harness this technology's full potential, and that doing so requires collaborative international action. Next year, the next generation of considerably more powerful models will be released and

participants identified a narrow window for clear, decisive, and committed action, to engage constructively, globally, and inclusively. They noted that the challenges posed by frontier AI could not be resolved at a single Summit, but that such discussions would set the foundation for realizing the ambitions of the Bletchley Declaration and into the next Summits, hosted by the Republic of Korea and France.[44]

While it is encouraging to see the UK, like the EU, take the lead in proposing a safe path forward for the world in the development of AI technologies, very little was said about the potential for existential risk of an SMG. And this is perhaps not surprising. Although many politicians are now starting to realize that Big Tech is rapidly racing to build an SMG, it would appear that the current concern for AI governance is focused on emerging forms of frontier ANI. It will be interesting to see how discussions evolve during the next two AI summits held in Korea and France as the imminent and impending emergence of AGI looms on the ever-advancing horizon.

As I am writing this book, it is important to remember that we are living through a unique time in history where many of the world's leaders are unfamiliar with the potential for harm from AI technologies, the power of which they have little or no comprehension. This is not to blame the politicians for their lack of foresight. They are incredibly busy people with enormous amounts of responsibilities and duties. Therefore, they cannot know anywhere near what experts within the AI fields commonly know and accept. That's why it is the job of AI risk and governance agencies to inform industry leaders, politicians, and especially the public on the potential for harm by such emerging new and transformative forms of AI technologies.

U.S. AI Bill of Rights

President Biden often says, "America is the only nation that can be defined by a single word: possibilities." The White House Office of Science and Technology Policy (OSTP) works to bring that idea to life by harnessing the power of science, technology, and innovation to achieve America's great aspirations.[45]

Arati Prabhakar, Ph.D., is director of the White House OSTP and assistant to the president for science and technology. In this capacity, Prabhakar is the president's chief advisor for science and technology, a member of the president's cabinet, and cochair of the President's Council of Advisors on Science and Technology (PCAST).[46]

OSTP's six policy teams work to advance critical administration science and technology priorities.

The Technology team works to advance technology and data to benefit all Americans. This includes leveraging technology and data to equitably deliver services, bringing technology and data expertise to federal policy formation and implementation, and ensuring that America continues to lead the world in values-driven technological research and innovation. The team works to harness the benefits of artificial intelligence (AI) for the American people while managing its risks, including through the National Artificial Intelligence Initiative Office (NAIIO), which advances and coordinates key federal initiatives and policy on AI. The Tech team also includes the U.S. Chief Data Scientist, who works to ensure that data science helps equitably tackle [the] nation's biggest challenges.[47]

To advance President Biden's vision, the White House Office of Science and Technology Policy has identified five principles that should guide the design, use, and deployment of automated systems to protect the American public in the age of artificial intelligence. The Blueprint for an AI Bill of Rights is a guide for a society that protects all people from these threats—and uses technologies in ways that reinforce our highest values. Responding to the experiences of the American public, and informed by insights from researchers, technologists, advocates, journalists, and policymakers, this framework is accompanied by *From Principles to Practice*—a handbook for anyone seeking to incorporate these protections into policy and practice, including detailed steps toward actualizing these principles in the technological design process. These principles help provide guidance whenever automated systems can meaningfully impact the public's rights, opportunities, or access to critical needs.[48]

FIVE PRINCIPLES THAT SHOULD GUIDE THE DESIGN, USE, AND DEPLOYMENT OF AUTOMATED SYSTEMS TO PROTECT THE AMERICAN PUBLIC IN THE AGE OF ARTIFICIAL INTELLIGENCE

1. **Safe and Effective Systems**

   The public have a right to be protected from unsafe or ineffective systems. Any automated systems that are developed must be done so with "consultation from diverse communities, stakeholders, and domain experts to identify concerns, risks, and potential impacts of the system." All such systems must undergo pre-deployment testing prior to deployment to the public. As well, potential and identified risks need to be identified and mitigated, with continuous monitoring that assures their safe and intended use.

2. Algorithmic Discrimination Protections

No members of the public should face discrimination caused by algorithms; systems must be designed and operated equitably. Algorithmic discrimination happens when automated systems unfairly treat people differently based on factors like their race, color, ethnicity, gender, religion, age, national origin, disability, veteran status, genetic information, or other legally protected categories.

3. Data Privacy

The public should be protected from data practices that are abusive. Additionally, the public should have agency over how their private and public data is used. Those who create and use automated systems must ask for permission and respect as much as possible personal choices about how data is collected, used, shared, and deleted.

4. Notice and Explanation

The public have a right to know when they are being subjected to an automated system and to understand the ways in which it impacts and affects them. All creators of automated systems must explain in clear language how their system works, the role it plays, and those responsible for its development and use.

5. Human Alternatives, Consideration, and Fallback

Where appropriate, the public must be given an option to opt out and be able to contact an individual who can respond quickly and efficiently to any problems that arise. As well, at any time, the public should be given the choice of opting out of using an automated system in favor of an actual human with which to communicate. In cases where humans are required by law, the public should be granted timely access—especially in cases where automated systems fail to comply.[49]

The Blueprint for an AI Bill of Rights offers this conclusion:

Considered together, the five principles and associated practices of the Blueprint for an AI Bill of Rights form an overlapping set of backstops against potential harms. This purposefully overlapping framework, when taken as a whole, forms a blueprint to help protect the public from harm. The measures taken to realize the vision set forward in this framework should be proportionate with the extent and nature of the harm, or risk of harm, to people's rights, opportunities, and access.[50]

THE U.S. EXECUTIVE ORDER ON AI

On October 30, 2023, two days before the UK Summit on AI, the White House released President Joe Biden's Executive Order on the Safe, Secure, and Trustworthy Development and Use of Artificial Intelligence.[51] The executive order is detailed and quite lengthy and considers many important and relevant factors regarding the governance of AI as the world continues to develop various forms of technologies.

> Today, President Biden is issuing a landmark Executive Order to ensure that America leads the way in seizing the promise and managing the risks of artificial intelligence (AI). The Executive Order establishes new standards for AI safety and security, protects Americans' privacy, advances equity and civil rights, stands up for consumers and workers, promotes innovation and competition, advances American leadership around the world, and more.[52]

The order goes on to state that it is directing "the most sweeping actions ever taken to protect Americans from the potential risks of AI systems."[53] This order aims to enhance safety and security in the rapidly advancing fields of artificial intelligence. Its key points are as follows:

Sharing Safety Information: Companies developing powerful AI systems must share safety test results and other critical information with the U.S. government. In other words, any company "developing any foundation model that poses a serious risk to national security, national economic security, or national public health and safety must notify the federal government when training the model, and must share the results of all red-team safety tests."[54] This type of transparency will not only garner trust but also will ensure that AI systems are safe, secure, and trustworthy before they are made public, particularly if they pose risks to national security or public safety.[55]

Setting Standards and Testing: The National Institute of Standards and Technology will establish rigorous standards for testing AI systems for safety using red-team testing to ensure safety before they are released to the public. "The Departments of Energy and Homeland Security will also address AI systems' threats to critical infrastructure, as well as chemical, biological, radiological, nuclear, and cybersecurity risks. Together, these are the most significant actions ever taken by any government to advance the field of AI safety."[56]

Protecting against Biological Risks: The establishment of strong standards will be developed to screen for dangerous biological materials created with the help of AI. Also known as "biological synthesis" or the creation of novel, new forms of biological agents, the ability or capacity to use AI to assist in making deadly biological materials is one of the greatest concerns in AI risk at this period in time. Therefore, any and all agencies funding life-science projects will require compliance with these standards to manage risks associated with AI.

Protect Americans by Combatting AI-Enabled Fraud: The Department of Commerce will develop guidance for content authentication through the development of standards and best practices to detect AI-generated content and authenticate official content. This includes labeling AI-generated content and ensuring that communications from the government are authentic through the use of "watermarking" to clearly label AI-generated content.

Establish an Advanced Cybersecurity Program: A program will be established to develop AI tools for identifying and fixing vulnerabilities in critical software, enhancing cybersecurity. This builds on existing efforts to address cybersecurity challenges using AI, making networks and software more secure.

Order the Development of a National Security Memorandum: A memorandum will be developed by the National Security Council and White House chief of staff to guide further actions on AI and security, ensuring that the military and intelligence community use AI safely, ethically, and effectively. It will also "direct actions to counter adversaries' military use of AI."[57]

Biden's executive order then goes on to outline how the U.S. federal government is going to protect the American people by assuring seven key factors and the "actions" needed to be taken to ensure their fulfillment.

The first key factor is **Protecting Americans' Privacy**:

> To better protect Americans' privacy, including from the risks posed by AI, the President calls on Congress to pass bipartisan data privacy legislation to protect all Americans, especially kids, and directs the following actions:
>
> - Protect Americans' privacy by prioritizing federal support for accelerating the development and use of privacy-preserving techniques. . . .

- Strengthen privacy-preserving research and technologies, such as cryptographic tools that preserve individuals' privacy, by funding a Research Coordination Network to advance rapid breakthroughs and development. . . .
- Evaluate how agencies collect and use commercially available information—including information they procure from data brokers—and strengthen privacy guidance for federal agencies to account for AI risks. . . .
- Develop guidelines for federal agencies to evaluate the effectiveness of privacy-preserving techniques, including those used in AI systems. These guidelines will advance agency efforts to protect Americans' data.[58]

The second factor involves **Advancing Equity and Civil Rights**. To ensure that equity and civil rights will be protected as AI advances, President Biden directs a number of actions such as the clear guidance of landlords, as well as federal benefits programs and contractors to prevent algorithms from worsening cases of discrimination. This can be accomplished through training and technical assistance from the Department of Justice and federal civil rights. Biden also wants to ensure that the criminal justice system is also onboard in its use of AI technologies for "sentencing, parole and probation, pretrial release and detention, risk assessments, surveillance, crime forecasting and predictive policing, and forensic analysis."[59]

With the third factor, Biden promises to **Stand Up for Consumers, Patients, and Students**. Biden wants to make better and cheaper forms of AI technologies available for everyone while limiting their harmful effects. He believes this can be done in two ways: by advancing the responsible use of AI in health care and by developing affordable life-saving drugs. As well, the Department of Health and Human Services should create a safety program to monitor and correct any harmful or unsafe health care practices that use AI. In reference to education, Biden wants teachers to utilize AI-enabled education tools—such as personalized tutoring—as a transformative resource.[60]

Biden considers how workers will be affected by AI as his fourth key factor: **Supporting Workers**. Biden's advisors have accurately advised him that AI will have a considerable impact on the workforce in some areas. Hence, in an effort to proactively mitigate these risks, Biden believes we must mitigate the harms of AI and maximize its benefits by anticipating and addressing the number of jobs that will be lost to AI, and to study and report on its impact on the labor market. This will provide a clearer path to identifying labor options available to those who face such disruption.[61]

The fifth key factor Biden considers is the **Promotion of Innovation and Competition**. In reference to encouraging innovation and the entrepreneurial spirit in businesses like start-ups, Biden urges Americans to lead the way by developing a National AI Research Resource that will inform both AI researchers and students to access key AI resources, data, and grants in important areas like climate change and health care. Biden also wants to promote an AI ecosystem that is fair, open, and competitive to entrepreneurs and developers by offering resources and technical assistance that can help small businesses monetize AI advancements while encouraging the Federal Trade Commission to monitor developments and exercise its authority. Biden believes this will also help attract and retain highly skilled immigrants and nonimmigrants with expertise in needed areas to stay in the United States by making visa criteria, interviews, and reviews more modernized and streamlined.[62]

Biden's order encourages the continuation of working with countries throughout the world and **Advancing American Leadership Abroad** to support "safe, secure, and trustworthy deployment and use of AI worldwide."[63] To ensure this, President Biden intends to expand bilateral, multilateral, and multistakeholder collaborations on AI, which will include the State and Commerce Departments that can collectively establish meaningful international frameworks that ensure the benefits of AI while simultaneously mitigating its risks and ensuring safety to all parties involved. Biden believes that this process will speed up the development and implementation of AI standards with international partners in an effort to responsibly solve global challenges and sustain development by ensuring that such emerging technologies are "safe, secure, trustworthy, and interoperable."[64]

The seventh and final key factor President Biden's executive order addresses is to **Ensure Responsible and Effective Government Use of AI**. President Biden promises to ensure responsible government deployment of AI and to also modernize federal AI infrastructure. To accomplish this, the president intends to establish clear standards as guidance to agencies that use AI in an effort to protect human rights, and to safely improve how AI is procured and deployed. He also wants to help agencies to purchase AI products cheaper, faster, and more effectively "through more rapid and efficient contracting." [65]

And finally, President Biden wishes to use the Office of Personnel Management, the U.S. Digital Service, the U.S. Digital Corps, and the Presidential Innovation Fellowship to rapidly accelerate the hiring of professionals in AI in a government-wide AI talent surge. These agencies will then provide essential training for employees at all levels and in all relevant fields.[66]

Although the actual executive order[67] is much longer, more dry, boring, and detailed, it is a very good start to addressing the urgent need for AI governance in the world. Overall, President Biden's executive order aims to address the most central hope of all those working in AI safety today: to determine and address potential risks associated with AI while harnessing its benefits for public safety, national security, and global harmony.

China's AI Governance

It was hopeful to see Chinese officials attend the UK Summit on AI in early November 2023. What nobody is saying too loudly right now is that the race to be the first to build a superintelligent machine god (SMG) includes some known players such as OpenAI, Microsoft, DeepMind, Meta, and Anthropic; but we have a limited idea as to how close China or other major countries—like Russia—might be in the race. Some have estimated China to be anywhere from six months to several years behind the United States in their AI developments. Nor do we know what any one company or country could or might do once AGI is achieved. Since "transparency" is considered to be one of the most overarching ethical principles and values that China and most other countries have agreed upon, what do we currently know about China's regulatory commitments?

> The Chinese government expects for AI to create $154.638 billion USD in annual revenue by 2030. China, however, is not just focused on the proliferation of AI and its innovative use cases; the country has also been silently leading the pack and making its mark on the AI regulatory landscape. In 2022, China passed and enforced three distinct regulatory measures on the national, regional and local levels. This momentum carried into 2023, where in January alone, China cracked down on deepfake and generative technology through national-level legislations.[68]

Of the three distinct regulatory measures put forward by China, the first is called "Internet Information Service Algorithmic Recommendation Management Provisions," which became an active regulation as of March 1, 2022.[69] It contains six chapters covering topics such as general provisions, regulation of information services, protection of users' rights and interests, oversight and management, legal responsibility, and supplemental provisions. Throughout the six chapters, we find thirty-five articles that address some familiar but important issues. Although there is not enough space to list them all here, it is interesting to see how they embrace similar desires to protect the interests of members of the People's Republic of China homeland first (as did Biden with the United States), but they also acknowledge the importance of observing high ethical standards in good faith. For example, the very first article under "General Provisions" states:

> These Provisions are drafted on the basis of the Cybersecurity Law of the PRC, the PRC Data Security Law, the Personal Information Protection Law of the PRC, the Measures on the Administration of Internet Information Services, and other relevant laws and administrative regulations, so as to regulate internet information services algorithmic recommendation activities, carry forward the Core Socialist Values, preserve national security and the societal public interest, protect the lawful rights and interests of citizens, legal persons, and other organizations, and promote the healthy and orderly development of internet information services.[70]

But by the fourth article, we see a much broader recommendation:

> The provision of algorithmic recommendation services shall abide by laws and regulations, observe social morality and ethics, abide by commercial and professional ethics, and respect the principles of fairness and justice, openness and transparency, science and reason, and sincerity and trustworthiness.[71]

This is quite similar to the recommendations of the European Union, United States, and United Kingdom. Since there is collective agreement by many of the world's most powerful nations on how we should proceed with the development of AI technologies, it would appear that we find ourselves back in our prisoner's dilemma whereby everyone will be better off as long as we agree upon and cooperate to a common set of ethical precepts and international laws.

Some of China's guidelines are quite specific and deal with the potential for sociotechnical issues to arise such as that found in article 8 of their chapter on regulating information services: "The providers of algorithmic recommendation services shall periodically check, assess, and verify algorithm mechanisms, models, data, and outcomes, and must not set up algorithmic models that violate laws and regulations, or go against ethics and morals, such as by inducing users to become addicted or spend too much."[72] It is interesting to see how China has considered the potential for manipulation with such forms of technology, which, by the way, have already been used in online gaming sites in the United Kingdom and other countries. From just one random search on a gambling website, I found the following:

The algorithm is able to tailor gaming experiences, bonuses and offers to match individual player profiles by gathering and processing a player's gaming information. The algorithm constantly improves its insights as more people interact with it, making for a lively and exciting result.[73]

It's interesting to see how "the algorithm" is spun as a good thing, which only gets better by constantly taking in more and more information about you—thus, constantly "enhancing" your gambling experience. The reality of the sociotechnical backlash of this form of AI, however, is that it also uses information about you to manipulate you into spending more than you can afford. In the UK, for example, we find that "the algorithm" can be downright deceitful according to Matt Zarb-Cousin, a leading gambling reformer in the UK as well as a recovering gambling addict. Zarb-Cousin maintains that large gambling operations extract enormous amounts of data from their users and then use it against them when they are at their most vulnerable:

Mark Zarb-Cousin successfully lobbied for stricter gambling regulations in Britain—limiting how betting companies advertise and how much gamblers can wager. He says the U.K.—where gambling's been legal for decades—offers a sobering glimpse into what he believes is a crisis headed straight toward the U.S.[74]

What Zarb-Cousin discovered was that the larger online gambling sites were using algorithms that did not have the gamblers' best interests in mind: "There's lots of opportunities to gamble in Britain. You assume it's safe. You don't realize how easy it is to get addicted to that stuff. . . . Addiction is intensified by how much the gambling companies know about each user."[75]

Recently, Zarb-Cousin used Britain's public information laws to gain access to data from the betting company Flutter—owner of FanDuel—to check the influence it had on its UK customers. He discovered that the algorithmic extraction of data of each user was being used to manipulate them to continue gambling:

So, about 93 different data points they had on this individual were when they bet, what offers worked, what inducements worked. On this particular one, he played slots for three to four days straight. They knew the life stage—the customer life stage he was at. So, "win back," they described it: so people that have given up gambling for a while, and they're trying to get them to come back.[76]

Since most of the gambling companies know when gamblers are at their most impulsive, and they have a great deal of information about them, it no longer makes the wager fair, according to Zarb-Cousin. It would appear that China has recognized this type of manipulation as a genuine threat to its citizens and is proactively attempting to address it before it can become a problem.

Other articles consider issues such as addressing the rights of seniors to be safe from online manipulation, maintaining the confidentiality of personal private information, and enforcing fines if violations to such articles occur.

The second distinction regulatory measure China has taken is called 'Provisions on the Administration of Deep Synthesis Internet Information Services."[77] This document, with its twenty-five articles spanning five chapters, covers some of the recommendations for local telecommunications and public security departments responsible for the oversight of deep synthesis services in the corresponding administrative regions and in accordance with their respective duties. For those of you wondering what is meant by "deep synthesis," it can be explained in the following way:

> AI-Deep Synthesis, also known as deepfake, is a technology that can generate or manipulate image, audio, or video content that appears realistic but is not authentic. It can be used for various purposes, such as entertainment, education, art, journalism, or political satire. However, it can also pose serious risks to individuals and society, such as identity theft, fraud, defamation, misinformation, or cyberbullying.[78]

In the document, various forms of "deep synthesis" are provided in the final chapter and defined as "the use of technologies such as deep learning and virtual reality, that use generative sequencing algorithms to create text, images, audio, video, virtual scenes, or other information."[79] By now, most of us will be quite familiar with how good AI technology is getting online. We know that such forms of video manipulation have, can, and will be used to spread false information. In China, there was a well-known case involving the manipulation of a famous influencer:

> Chinese influencer "CaroLailai" discovered her face was swapped with a porn actress in a viral video. She reported the illegal industry chain behind deepfake technology to the police. This case reminds people of ZAO, a deepfake app that sparked major privacy concerns in China, the Ministry of Industry and Information Technology (MIIT) ordered its removal from the app store in 2019. Similar cases still happen, some of which involve

criminals using deepfake technology to scam victims into transferring money or disclosing personal information.[80]

Due to the potential for harm that such AI-generated deepfake images could generate, China has decided to proactively address it and legislate for its control. As with the United States, European Union, and the United Kingdom, China has put forward the need to identify the use of AI-generated images and videos so the public can fairly and transparently know they have been created using such methods. For example, in chapter 3, "Data and Technical Management Specifications," article 17 states: "Where deep synthesis service providers provide the following deep synthesis services which might cause confusion or mislead the public, they shall make a conspicuous label in a reasonable position or location on information content they generate or edit, alerting the public of the deep synthesis generation." As we have seen earlier, this method of identifying or "watermarking" AI-generated content is generally a good and widely accepted way of informing the public. Should such fraudulent activity be used in any criminal activities, the resulting penalties will hopefully match the severity of the crime: "Where violations of public security are constituted, the public security organs are to give public security administrative sanctions in accordance with law; where a crime is constituted, criminal responsibility is pursued in accordance with law."[81] These rules took effect January 10, 2023.

The third and final distinct regulatory measure on the national, regional, and local levels of AI governance is called "Interim Measures for the Management of Generative Artificial Intelligence."[82] This five-chapter document with twenty-four articles covers measures that "apply to the use of generative AI technologies to provide services to the public in the [mainland] PRC for the generation of text, images, audio, video, or other content (hereinafter generative AI services)."[83] This document has a little more of a nationalist tone to it. See if you can spot it:

> The provision and use of generative AI services shall comply with the requirements of laws and administrative regulations, respect social mores, ethics, and morality, and obey the following provisions: (1) Uphold the Core Socialist Values; content that is prohibited by laws and administrative regulations such as that inciting subversion of national sovereignty or the overturn of the socialist system, endangering national security and interests or harming the nation's image, inciting separatism or undermining national unity and social stability, advocating terrorism or extremism, promoting ethnic hatred and ethnic discrimination, violence and obscenity, as well as fake and harmful information.[84]

It's interesting to see the reference to ethnic hatred and discrimination. Most of the world is somewhat familiar with human rights abuses in China, which, coming from the U.S. Department of State, includes—but is by no means limited to—the following:

- Repression in Xinjiang[85]
- Fear of arbitrary arrest
- Religious freedom abuses
- Stifling freedom of expression
- Forced labor
- Assault on Hong Kong's autonomy
- Severe restrictions in Tibet[86]

So . . . in principle, China's intentions sound great. But in application, they may not be as forthcoming. And then, of course, there's TikTok. How much control did Bytedance have over the massive amounts of data on its platform? And how much data went to the PRC? We may never know. But somebody does.

There is also mention of respecting intellectual rights, which has an element of humor because China is generally the place where knock-off items are born and widely sold. Aside from these more obvious discretions, the spirit of the document puts forward commonly shared values such as to "encourage the innovative application of generative AI technology in each industry and field; generate exceptional content that is positive, healthy, and uplifting; and explore the optimization of usage scenarios in building an application ecosystem."[87]

As well, emphasis is placed on supporting "industry associations, enterprises, education and research institutions, public cultural bodies, and relevant professional bodies, etc. to coordinate in areas such as innovation in generative AI technology, the establishment of data resources, applications, and risk prevention."[88] So it would appear, in theory at least, that some measures are being addressed that could lead to a more fair and just governance of such new and emerging AI technologies.

It is important to see some of the details behind the governance of AI between different countries and nation-states to better understand the common values and principles that each shares and agrees to abide by. Now, *how* each country will act on these principles and values is another matter entirely. In a perfect world, every country would understand the prisoner's dilemma that we're all in (i.e., as long as everyone cooperates, everyone will do better). The benefit of one company, or one country, will cause all boats to rise, and therefore all should want success and alignment with our race to build our SMG. And if we're all good Hobbesians, we will all abide by our world social contract, which has a strong sovereign in place who, if necessary, will quickly and effectively exact fitting consequences for the cheater or cheaters who want more than their fair share. But of course, it would be naïve to think that things will go this smoothly. We can certainly hope and aim for it. And perhaps humans will come to see the value in shared interests and benefits. These are the hopeful utopias of science fiction. To what degree and extent we move ourselves closer to utopia or dystopia with the increasing race to build an SMG is left for us to decide.

What the Experts Are Saying about AI Governance

All right, then, so where does all this leave us in terms of moving forward with AI governance? So far, we've seen a lot of theory; but how, exactly, will all of it play out in the real world? In conversation with my colleague at Convergence Analysis, Justin Bullock, he mentions that there are essentially two overarching approaches in dealing with AI governance:

1. **Break down efforts into three categories—legislation, institutions, and levers**

    A. Consider what forms of *legislation* are currently being developed in the world. Consider other areas such as nuclear material and armament governance as good indicators for emulating. Organizations like the Center for AI Safety,[89] Future of Life Institute,[90] and our team in AI governance at Convergence[91] are tracking this category.

    B. What kinds of *institutions* might we need to govern AI? It's a global problem, so people are imagining multilateral global institutions. Scholars like Matthijs Maas have done extensive work producing a comprehensive report on the difficulties involved in advanced AI governance, which he defines as "the study and shaping of local and global governance systems—including norms, policies, laws, processes, and institutions—that affect the research, development, deployment, and use of existing and future AI systems, in ways that help the world choose the role of advanced AI systems in its future, and navigate the transition to that world."[92]

C. What *tools/levers* could be used to control the ability to use transformative and powerful AI for harmful purposes? Bullock believes there are three main levers for AI governance:

a. Licensing
b. Insurance, strict liability
c. Compute governance[93]

We shall consider some of these a bit later in the chapter.

2. Broader approach: Begin with an ambitious idea of what is needed and work backwards

This approach considers such questions as the following: What academic fields might contribute to how we control AI and how AI might influence other things? What are, and should be, the components of a regulatory framework? For example, what is AI likely to do to markets, and what tools might we use to respond to that? How might AI tools change the distribution of cultural, political, and military power across major nation-state players?

When I asked Justin Bullock how he thought we might be able to guarantee that the ethical principles agreed upon by so many countries—such as transparency, reliability, and accountability—will be respected by all producers of AI, he suggested the idea that we should have an international treaty that enforces some upper limit on computing, perhaps using licensing and limiting models that have been trained with more than a specific threshold of computing power. Bullock believes that, when it comes right down to it, there are really two main levers for controlling AI development: computing power (or "compute" for short) and data. But how, then, should we determine governance over such levers when most legislation goes from municipal or local, to state or province, to national, and to international levels? In other words, do we know what's going on at these four levels of government? Bullock pointed out that there's a lot of variation throughout the world, even in the West, about how much authority cities, counties, states, and nations have. In a normal domain, he said, a few different things can happen:

In the United States, for example, the federal government sets the broad standards, and has the money, and works with states to implement enforcement. Then local governments are on the ground, and give some oversight at the county and up to state levels; that's the model for example, for transportation. But this is *not* the case for nuclear energy or biological R&D work, because of the risks involved. This is generally because of the large associated externalities. In other words, if it can affect the whole nation, it follows that you have to work on it at a national level; if it's going to affect the entire world, then work on it at the international level. The governance matches the scope of the risk. Therefore we would imagine that if people were taking AI seriously, we would have an international governing body that works with nation states to oversee stuff. This is the rational thing to do.[94]

Ah, reason—that cold, precise, virtuous capacity we all value so much but rarely agree upon when put into practice. My biases may be showing, but Bullock appears to be making sound, common, and rational sense here. Just as Carl Sagan once said: "Extraordinary claims require extraordinary evidence,"[95] so too would it follow that, the greater the likelihood for higher amounts of damage and destruction on larger scales, the more we, as a species, need to figure out how we're going to pull this off on a global scale. There are many in the AI biz who are as confident as Bullock in believing that we are headed in this direction. He believes that, currently, there is a lot of momentum among different sectors for this kind of top-down framework, but it depends on the use cases. Bullock believes the EU AI Act is a good template of what this might look like, because it implements both binding laws and inspection agencies.

For the record, there has been a lot of talk for years about how the global AI communities throughout all countries and nation-states might form an international governing body similar in scope and reach to the International Atomic Energy Agency (IAEA), with the most powerful AI models having to be registered with this hypothetical agency as if you were handling nuclear energy. "Or," says Bullock, "something like the NSF; not as global or top down, but you have to be a part of the consortium to access a given upper limit of compute."[96] In terms of policing, Bullock believes that, in the not-too-distant future, most of the compute will be controlled by a few of the wealthiest countries, which could bilaterally agree among each other and then regulate internationally from there.

When we considered the power differentiation between developed and less developed countries of the world, Bullock stated:

AGI will be reached by scaling of current methods—the U.S. and China are just so far ahead, so the chance of a poor country having a major breakthrough is unlikely. What is more likely is that the gap widens, because it requires so much money to train, gather data and attract talent, so the cutting edge goes further beyond what anybody else can develop. This stuff will concentrate in the wealthiest countries and companies, like the internet did.[97]

When I asked Bullock if he thought that some Big Tech corporations will have enough capital to simply defy governments and "go rogue" in their race to build an SMG, he said:

> In the modern world, sovereignty is better thought of as states and markets, not states *or* markets. The combination of governments and corporations working in partnership is what has domination of different areas of society (although it's unclear how AI might change the power dynamics). The state-corporation relationships will look more like they do in space. For example, NASA provides funding to work with private companies to keep the cutting edge, but most of it is done in private companies. In the way that Elon Musk had to publicly apologize for smoking weed to get money from NASA, there will be mutual reinforcement where governments and corporations need one another. If Altman wanted to take OpenAI to his own neutral island it would incur huge costs. So it's far more likely they will keep playing the game together.[98]

But in terms of governing the actual control of an SMG, are there currently protocols that would guarantee that we can achieve an ACCORD (Alignment, Control, Containment, or Danger)? To this, Bullock pointed out the work of some of his colleagues at Anthropic who try to measure how far an LLM will diverge from its function. Anthropic believes such powerful forms of AI can be trained under a kind of "constitution," which would make it very hard for it to deviate from its ethical precepts and principles. Anthropic maintains that

> [h]uman feedback can prevent overtly harmful utterances in conversational models, but may not automatically mitigate subtle problematic behaviors such as a stated desire for self-preservation or power. Constitutional AI offers an alternative, replacing human feedback with feedback from AI models conditioned only on a list of written principles. We find this approach effectively prevents the expression of such behaviors. The success of simple principles motivates us to ask: can models learn general ethical behaviors from only a single written principle? To test this, we run experiments using a principle roughly stated as "do what's best for humanity."

We find that the largest dialogue models can generalize from this short constitution, resulting in harmless assistants with no stated interest in specific motivations like power. A general principle may thus partially avoid the need for a long list of constitutions targeting potentially harmful behaviors. However, more detailed constitutions still improve fine-grained control over specific types of harms. This suggests both general and specific principles have value for steering AI safely.[99]

While it is heartening to see the attempt being made to keep powerful forms of AI within the moral guardrails we give it, I'm not totally confident in leaving it up to an SMG to "know" what is "best for humanity." Would this be an LLM trained on the majority of the world's information? Would it be boxed and therefore limited in what constitutes what is "best" for us? What guarantees are there that misalignment has been thwarted? Or at least, largely controlled or contained? What's best for humanity if you are American? Canadian? Australian? Live in a democracy? Live in an autocracy? What if you're a drug lord? A white supremacist? A racist? A Christian, Jew, Muslim, or atheist? There are countless biases that go into what might constitute what is "best" for humanity. How would this form of AI be able to determine and "know" what is best for all of us? Don't get me wrong; having ethical guardrails in the form of unequivocally unbreachable constitutional commands sounds like it may work on some levels. But I think we're going to need more than just a single precept.

Others in the AI biz, like Stuart Russell, believe that we can govern and control an emerging SMG by giving it not one, but three principles that implicitly reflect what he calls "human-compatible AI": "I'm trying to re-define AI, to get away from this classical notion of machines that intelligently pursue objectives."[100] In developing a theory of beneficial machines, which aims to provide a provable, theoretical fix for creating aligned AI, Russell posits three principles or laws as a guide for AI researchers that he argues will lead to creating aligned AI:

1. *The AI's only objective is to maximize the realization of human values.*

   This is an altruistic value in which the AI does not care at all about its own self-preservation (as in Asimov's Three Laws) but *only* for human values defined by Russell as "whatever it is that the human would prefer their life to be like."[101]

2. *The AI is initially uncertain about what those values are.*

This is a "law of humility," according to Russell, in which the AI is asked to maximize human values of which it is ignorant. This avoids the problem of a single-minded pursuit of an objective like the maximizing misalignment found in Bostrom's paperclip thought experiment.

3. *Human behavior provides information about human values.*

Russell believes that, in order for an SMG to be useful to us, it has to have some idea of what we want. So it obtains this information by observing actual human choices: "So our own choices reveal information about what it is we prefer our life to be like."[102]

These are thought-provoking principles, but I'm a bit curious about what might happen if such an AI system were to learn by simply observing "whatever it is that the human would prefer their life to be like" and our vastly differing choices reflecting those preferences. For example, what might happen if the robot observed a really morally corrupt group that treats its members with distrust, hatred, and violence? Would it learn to act in similarly despicable ways? But we don't even have to go that far. Just imagine all the sorts of nasty things humans do to each other and how they would prefer their lives to be better if it meant the suffering of others. Russell acknowledges that challenges still remain around, say, how such a machine might make determinations of such nuanced human harmful behaviors as passive aggression, rumor spreading, and sarcasm.

In reference to how these principles might be used to control or contain an AI involving the need to shut it down by hitting the kill switch, Russell outlines a basic misalignment scenario:

We have [a PR2 robot] in our lab, and it has a big red "off" switch right on the back. The question is: Is it going to let you switch it off? If we do it the classical way, we give it the objective of, "Fetch the coffee, I must fetch the coffee, I can't fetch the coffee if I'm dead," so obviously the PR2 has been listening to my talk, and so it says, therefore, "I must disable my 'off' switch, and probably taser all the other people in Starbucks who might interfere with me."[103]

Russell believes this scenario represents the classic type of inevitable behavior that an AI will take *if* we give it a concrete, definite objective. But if

252 BUILDING A GOD

it has no value system, then it won't reason in ways that could cause it to misalign:

> So what happens if the machine is uncertain about the objective? Well, it reasons in a different way. It says, "Okay, the human might switch me off, but only if I'm doing something wrong. Well, I don't really know what wrong is, but I know that I don't want to do it." So that's the first and second principles right there. "So I should let the human switch me off." And in fact you can calculate the incentive that the robot has to allow the human to switch it off, and it's directly tied to the degree of uncertainty about the underlying objective.[104]

Russell believes that once the AI is switched off, the third principle will be observed and realized by the AI (i.e., "human behavior provides information about human values"). Therefore, turning the AI off teaches it about what actions it should and should not do. But does this not assume that it would somehow have to "know" that being turned off is negatively valued? It's possible to imagine that the AI can infer that it was turned off because it was trying to do something humans negatively value, and therefore, in the future, it ought not to do that action again. Russell also believes that he and his team can mathematically prove a theorem that states that this type of robot is "provably beneficial to the human."[105]

Although Russell is not yet able to inform us about how he intends to keep a superintelligent machine god ignorant and subservient with his three principles, it's an interesting approach to controlling an SMG by keeping it ignorant of its objectives and human values. But I'm still a bit concerned about the power and level of intelligence of an SMG and whether such a being will still comply with them should it develop a level of self-awareness or consciousness. For once AGI is reached, we have no idea how it will behave. It may come to understand its own being in relation to ours and decide to do nothing but what's best for humanity and other species as well. It might stay ignorant and comply perfectly well with Russell's three principles. But it might not. It may think that its superintelligence warrants its decision to live by its own ethical precepts and principles. And maybe these align with ours; and maybe they don't. But I am a bit concerned whether any such being can be controlled or contained by Russell's principles—or perhaps any principles, whatsoever. And such misaligning behavior could result—whether the SMG is sentient and conscious or not.*

---

* Russell has since added to his alignment principles the idea of "assistance games," which demonstrate how an AI system could advance its learning objectives through a series of nudges, hints, or

And then there are people like nanotechnology science pioneer Eric Drexler, who believe that, as we develop more and more powerful forms of AI, we can keep them isolated to specific tasks:

> Intelligent systems optimized to perform bounded tasks (in particular, episodic tasks with a bounded time horizon) need not be agents with open-ended goals that call for self-preservation, cognitive enhancement, resource acquisition, and so on; by Bostrom's orthogonality thesis, this holds true regardless of the level of intelligence applied to those tasks.[106]

Drexler envisions a future where AI technologies are aligned and controlled because we simply never allow them to become superintelligent "agents." Instead, we build them to become superintelligent "services."

> In contrast to unprecedented-breakthrough models that postulate runaway self-transforming agents, prospects for the incremental emergence of diverse, high-level AI capabilities promise broad, safety-relevant experience with problematic (yet not catastrophic) AI behaviors. Safety guidelines can begin by codifying current safe practices, which include training and re-training diverse architectures while observing and studying surprising behaviors. The development of diverse, high-level AI services also offers opportunities for safety-relevant differential technology development, including the development of common-sense predictive models of human concerns that can be applied to improve the value and safety of AI services and AI agents.[107]

Drexler is attempting to "reframe" Nick Bostrom's idea of the potential threat of an SMG. Instead of putting all of our efforts into making a singular superintelligent machine god, let's instead make a bunch of superintelligent machine angels (SMAs) that are extremely good at what their very narrow and focused parameters allow them to be. So, if it's language translation you want, let's build a superintelligent AI translator that can never become an agent. If you need to know the most efficient way to travel across a country given a large number of variables, let the AI Mapper get you there with optimal speed and efficiency. Drexler believes if we can keep the scope and range of such technologies narrow enough, we can control them. And this may, in fact, turn out to be the case. In fact, we're already starting to see this trend happening currently in technologies like radiological diagnostics such as X-rays, MRIs, and CT scans. But this doesn't mean that others aren't going to continue the race toward

---

clues indicating the direction it should take.

the building of an SMG. Superintelligent machine angels (SMAs) are an interesting idea and will develop quickly and keep us amused on the road to building our SMG. But we will not stop the pursuit of building such a god because it is irresistibly in our nature to want to see "what's going to happen next."

The final expert view we will consider is that of Robert Trager, director of the Oxford Martin AI Governance Initiative. In an interview with Trager on my podcast *All Thinks Considered*, I asked him about what might happen if hostile countries like Russia, Iran, and North Korea were able to develop AI to the point of building an SMG. He said:

> I think this is a really interesting question. So one question is, you know, what are the risks that are associated with these kinds of technologies? And there's not agreement on that. We don't have everybody saying, "Oh, yes, if you crossed this particular line, then something really bad is going to happen" . . . in a way, that's a good situation because nobody wants disaster and nobody goes over that line. But the, I think the more likely situation is that it's more like a gradient descent, you know, a sort of a slippery slope. So it's not just disaster, when you go across a line, it's, well, when you go down a path, and it gets a little more risky. And that's really worrying, because then you have these competitive dynamics between all these different nations, whereby one says, "Oh, well, I'm sure my adversaries are going to go a little bit down this path to get an advantage. So I better go a little bit down this path." That doesn't mean that it's a race to the bottom, but it means there's a certain amount of risk taking there. And that's what I worry about.[108]

I asked Trager about his thoughts regarding the general governance of AI, such as whether or not he thought we should be putting ethical and legal safeguards up. Or should we be monitoring success rates? Or should AI developers have to report when they come across new developments that might be indicating that we're getting close to AGI? His response was to the point:

> Yeah, absolutely. I think we need all those things. And I think governments have to be involved. It should not be a set of folks in a private company that are making these decisions that affect not just everybody in one state, but people around the world. And so I think we need governance policies at the national level, but also at the international level. And that's really complicated to think about.[109]

It would seem that Trager is echoing what a lot of people in the AI safety biz want: policies at local, national, and international levels that will allow everyone to benefit while limiting the potential risks and harms of AI. One of the last questions I asked Trager was what he thought the general public might do to feel more empowered in dealing with AI developments. His response was optimistic:

> Oh, that's such a good question; I think we need to act collectively. We need social movements that will help people to feel empowered, where they can feel like they're a part of a movement, and they're getting together with others, and they're having a voice and things. You know, I think after World War II, for instance, there was the UN organization, not just the UN, but there was this UN organization of people that got together and felt like they were sort of participating in international governance and helping to kind of think about a New World Order, which hopefully would have less conflict. So I think those organizations play a big role. I think those sorts of social movements that collectively helped to give people a voice are important. They're important to people, psychologically; they're important in terms of the impact that they have. But they're also important just because they really do allow people to have a voice collectively.[110]

As we look further into the future of AI in the next chapter, we will consider in greater detail how the public can become more involved and, hence, more empowered in taking part in the discussion and direction of AI technologies.

Pushback against Regulation

Regardless of how little or how much regulation—government or otherwise—is applied to AI tech development, there will always be those who bemoan its necessity. For those who believe that AI regulators are overreaching on safety issues, there will always be organizations like the so-called anti-doomer DC nonprofit Alliance for the Future (AFTF):

> Some technologists have launched Alliance for the Future (AFTF), a DC-based nonprofit organization meant to fight AI safety forces linked to regulatory capture and perceived overreach. "AFTF works to inform the media, lawmakers, and other interested parties about the incredible benefits AI can bring to humanity. We will oppose stagnation and advocate for the benefits of technological progress in the political arena," the group writes in a statement. "Escalating panic and reckless regulation around artificial intelligence will cause more harm than benefit. AFTF was founded to

be the voice of ordinary users, builders, and founders, who want the basic freedom to use machine learning in their day to day lives."[111]

Torn from a similar page of Marc Andreessen, the AFTF sounds very much like they are on the "Drill baby, drill!" team. Jack Clark, cofounder of Anthropic and member of the U.S. government's National AI Advisory Committee, believes that "every action in policy creates a counter reaction: AFTF exists because a load of people affiliated with the AI safety community have lobbied in DC for ideas like needing licenses to develop AI systems, and other ideas that have generally been perceived as overreach."[112] This is why Clark believes organizations like AFTF form in the first place. In a Newtonian sense, in which for every action or force in nature there is an equal and opposite reaction, Clark states the we should keep in mind "that well intentioned policy is still a thing that exists in politics—and in politics forces always generate counter-forces."[113] There will always be an art to politics in finding the right balance between autonomy and paternalism; or, in other words, between liberty—how much freedom an individual or group can exercise in developing new AI technologies versus how much control or regulation (like a parent) should governments exercise in controlling them. There are no easy answers, and we can only hope that this delicate balancing act is carried out with the proper guidance and advice from those most knowledgeable in the fields of AI.

Do Any Silver Bullet Solutions Exist?

One of the questions I ask most frequently in the AI biz is whether or not there are single or relatively simple solutions to the race to build an SMG. Is there some technofix that would allow us to align, control, or contain an SMG? We saw in the last chapter a number of different proposals ranging from "boxing" or "closing the gate" on such advanced technologies before they become too advanced—a sort of Drexlerian solution whereby we make the AI really intelligent; just not generally superintelligent. Perhaps this might even define a new "type" of AI called super narrow intelligence (SNI). We also considered Steve Omohundro and Max Tegmark's provably safe systems whereby we can take a peek under the hood to ensure compliance. Or perhaps Russell's three principles may serve useful in this endeavor.

But is there some single act a governing body could do that would be relatively simple but effective in controlling some of the damage an SMG could do? These are often referred to as "silver bullet" solutions because they're based on the mythical horror movie trope that it just takes one silver bullet through the heart of a werewolf to kill it. Well, when we consider agencies like the International Atomic Energy Agency (IAEA), we see how it monitors important activities related to nuclear energy and atomic weaponry. By knowing who is doing what with uranium and other related atomic elements, the IAEA knows fairly well what is currently happening in the world with respect to atomic power and weaponry. The IAEA knows that there is only so much atomic material in the world. And if they can keep track of it and who has it, they have a pretty good handle on what's going down throughout the world. In a similar fashion, could an international AI agency do the same thing? Bullock mentioned levers for compute and data. If such an agency could control computing power, it could theoretically know about global developments of an SMG and impose limits and penalties when specific developers were not complying. But realistically, only a few lucky organizations or countries will be able to generate enough resources to attain the computing power to build such a god. And what does the computing power—or compute—subsist in? Microprocessing chips. Bullock and others believe that, if governments or other agencies could control the distribution of these chips, they could control the harmful effects of those who would use such power for corrupt or destructive purposes. But how would this work? I'm glad you asked.

One of our researchers at Convergence, Deric Cheng, has done a critically necessary deep dive into the AI chip game. And here's what he found:

> The training and inference of frontier AI models require massive numbers of advanced semiconductor chips, which themselves require highly advanced infrastructure to build at scale. The production of these cutting-edge semiconductor chips is currently a major bottleneck for the AI industry. Therefore, regulating chip production and purchasing is a powerful and direct way to control the development and usage of frontier AI models.[114]

Cheng believes that a powerful country like the United States is currently leading the way by limiting the ability for potentially hostile countries to access specific AI chips:

> The U.S. executive branch is likely to be interested in implementing an AI chip registration policy in the near future due to its geopolitical AI race with China. As evidenced by its stringent and comprehensive export controls limiting Chinese access to high-end semiconductors, the U.S. executive branch (specifically the Biden administration) has displayed a willingness to take decisive and far-reaching action to slow China's AI

progress. Over the past two years, it has repeatedly increased the breadth of its export controls to improve enforcement, including strengthening restrictions on the types of chips permitted to export and restricting access for certain countries in the Middle East. In its quest to reduce chip smuggling to Chinese organizations, a blanket high-end AI chip registration policy is one of the next most likely enforcement mechanisms.[115]

Cheng believes that a semiconductor chip registration policy would help the U.S. government pursue its AI safety interests by immediately allowing it to track the relative distribution and possession of high-end AI chips across different organizations and uncover organizations that have substantial resources in AI chips but may have previously flown under the radar.[116] But it needn't stop there. Such a policy could, according to Cheng, lay the groundwork for the United States to implement future regulations based on chip monitoring that may have a more active and substantial impact on AI safety. This could include the reduction of the rate of AI scaling due to competition and the restriction of chip purchasing power to those companies who do not comply with U.S. regulations.

It is of note to see that, even though China was well represented at the UK Summit on AI, and China's proposed policies are based on the same ethical precepts and principles found in the proposals of the European Union, United Kingdom, and the United States, when it comes right down to it, prohibitive action is already being taken against them. And why is that? The answer is obvious. China is poised to knock off the United States to become the next potential world leader. Should they gain any type of advantage through AI, this may speed up the process significantly. Hence, it is very much in the interest of the United States (and perhaps other countries as well) to limit or minimize China's ability to excel in AI technological development. Stop and think for a moment, dear reader, what would China do if they were the first to build an SMG? What would Russia and Putin do? What would Iran do? What would North Korea do? What would Hamas do? Or Israel? Fear, ignorance, and hatred are a dangerous combination to guide the direction of such power. Is there a Frodo among our world leaders who will resist the temptation to use the AI ring for evil purposes? We shall see.

Conclusion

Can we stop the threat of existential risk from an SMG by enforcing one or a series of laws that would disallow the eventuality of it either going rogue or, far more likely, falling into the hands of companies or countries that wish to use it for evil or harmful purposes? Perhaps; perhaps not. At this point, we just don't know. But try, we must. For the purpose of organizations like Convergence Analysis and several others throughout the world is to work toward ensuring that such an SMG never goes rogue and never falls into the wrong hands. The clock is ticking.

The Future of AI   6

Having explored the undoubtedly exciting advantages artificial intelligence (AI) could offer society, it is imperative that we address the potential societal chaos, catastrophic consequences, and existential threats advanced forms of AI may pose to humanity. We have encountered existential threats in the past that we continue to live with today; the Doomsday Clock ominously ticks toward the midnight of nuclear Armageddon, currently positioned at its closest ever due to the conflict in Europe (figure 6.1).

While we lack an emotionally resonant metaphor akin to the Doomsday Clock for a superintelligent machine god (SMG), the peril to humanity's existence may be equally genuine and unexpectedly imminent. And this is because we really don't know what time it is relative to reaching this goal. It is simply our biggest *known unknown* as Big Tech companies and political powers race at breakneck speed to be the first to accomplish this greatest

Figure 6.1.   The Doomsday Clock. *Wikimedia Commons.*

of all human feats. There is no clear consensus on exactly when this will be achieved. My colleague, Zershaaneh Qureshi, has recently attempted to clarify the variable terrain of artificial general intelligence (AGI) time line predictions. One key observation from her work is that many experts across different domains have moved their AGI predictions considerably closer in recent years.[1] This is due in large part to the recent advent of new AI technologies like large language models (LLMs), such as ChatGPT, Claude, and others, which have enormous access to data and possess huge amounts of computing power. As noted throughout this book, such LLMs have pushed up the AGI time line predictions considerably. Indeed, many of my colleagues and I believed just a few years ago that the development of AGI was perhaps forty years away. Now, we suspect that it could arrive in less than five years—perhaps ten years at the most.

Since we are knowingly ignorant as to when someone will succeed in building an SMG, it would be wise to prepare now, rather than later, for this eventuality. What we do know for sure is that someone, at some point in the not-too-distant future, is going to succeed in building such a superintelligent machine god. That means it is up to industry leaders, AI experts, and politicians—but especially us—to decide how we should proceed toward this inevitable goal.

As we noted earlier, in 2023, Sam Altman, CEO of OpenAI, went on a world tour and talked to politicians with an open plea for greater government regulation on AI technologies—up to and including those that may pose an existential risk to humanity.

> We have tried to be very clear about the magnitude of the risks here. My biggest fear is that we—the technology or industry—cause significant harm to the world. I think if this technology goes wrong, it can go quite wrong and we want to work with the government to prevent that.[2]

But others believe that, while it is admirable for such Big Tech companies to seek out AI risk governance, it is another matter entirely whether or to what degree such companies will comply.

> Unless there is external pressure to do something different, companies are not just going to self-regulate. . . . We need regulation and we need something better than just a profit motive. . . . It feels like a gold rush. . . . In fact, it is a gold rush. And a lot of the people who are making money are not the people actually in the midst of it. But it's humans who decide whether all this should be done or not. We should remember that we have the agency to do that.[3]

Other industry leaders of this gold rush such as Bill Gates, Elon Musk, Demis Hassabis, Anthropic CEO Dario Amodei, and executives from Amazon, Alphabet, and Meta have all expressed considerable interest in having government overviews of their emerging AI systems.[4] But though there has been agreement for the need for government regulation, it's not slowing down any of the Big Tech giants or world powers from going full-speed ahead in their attempts to be the first to build an SMG. And this should be a sounding bell and wake-up call for us, the unwashed masses. How does our future look as more and more advances continue in all areas of AI development? And what can we do to exert control of our collective future? Let's look at what some of the best minds in the AI biz have to say about it.

What Experts Are Saying about the Future of AI

As you might imagine, there are quite a few predictions about what the future of AI will look like. Some deal with how, precisely, an SMG will come to be built; others warn of more specific societal effects.

No SMG without Embodiment

Some AI experts believe that all the computing power and scaling up in the world is not going to build a sufficiently powerful AGI form of a superintelligent machine god (SMG). There is a school of thought that maintains that the "god-in-a-box" scenario is simply not going to happen because it requires one very important missing aspect: embodiment. If you think about it, animals as well as humans learn about their environments by physically navigating through them. It's one thing to know all of the true facts about what it means to ride a bicycle; but you can't really say you know "how" to ride one until you've physically sat on one and began pedaling and steering. This distinguishes the difference between propositional knowledge (i.e., knowing facts about something) and procedural knowledge (i.e., knowing "how" to do something). This leads us to question the so-called scaling hypothesis, in other words, that AI will continue to improve its abilities and performance forever with more and more compute and data until it transitions from artificial narrow intelligence (ANI) to artificial general intelligence (AGI) to artificial super intelligence (ASI). Whether or not advanced AI scaling up can become an SMG just by possessing propositional knowledge remains to be seen, but it is the central reason for concern at this point in our history. There's a school of thought in AI that maintains that procedural knowledge is what's going to provide that added dimension, which is considered both necessary and sufficient for building an SMG.

> ChatGPT doesn't have a body—but some AI researchers think "embodied cognition" is a necessary ingredient to achieve the field's holy grail of artificial general intelligence, or AGI. Others, including ChatGPT creator OpenAI, are betting all they need to reach AGI is to keep scaling up today's large language models with more data and more computational power. Language, reasoning and other abstract skills tend to get the most credit for human intelligence. But gaining knowledge of how the world works by walking, crawling, swimming or flying through it is an important building block of all animal intelligence. A group of prominent AI researchers last year advocated for an "embodied Turing test" to shift the focus away from AI mastering games and language, which are "well-developed or uniquely human, to those capabilities—inherited from over 500 million years of evolution—that are shared with all animals."[5]

This school of thought believes that both the human body and brain evolved simultaneously together, and as such, intelligence doesn't simply reside in the brain. Instead, there exists a type of mechanical intelligence that results from "embodied cognition," which "helps animals understand how the world works—by experiencing it."[6]

> "There's an argument to be made that biological systems learn from interacting with the world," says Jochen Triesch of the Frankfurt Institute for Advanced Studies. "Most machine learning systems today learn by basically passively absorbing large data sets, whether it is video or images or captioned images," Triesch says. Learning through interaction with the world is something "really essential that most of the machine learning community is right now completely missing."[7]

Triesch and his research team have developed a virtual human model of an eighteen-month-old child with five-fingered hands named MIMo. Its virtual body can sense its surroundings through its binocular vision, proprioception (the ability to sense your body's movement, action, and location), and a full body composed of virtual skin. The team believes it is the tactile nature of the skin that will assist in the embodied cognition of MIMo.

While researchers like Triesch and his colleagues work on "embodied cognition," there's another school of thought that's working on a little something called "distributed intelligence." A company called VersesAI, which is led by chief scientist Karl Friston, believes the road to building an SMG requires a different tack than the scaling-up approach of today's generative AI models. Friston and his team of researchers have gained inspiration by looking at how natural ecosystems work.

> Digital intelligence based on a web of intelligent agents is potentially cheaper, more environmentally sustainable, and more geopolitically defensible than one vast system trained on billions of data points. AGI is the industry's grail—a human level of artificial intelligence that can reason and learn in new ways—and Verses' founders say today's most advanced large language models, like OpenAI's GPT-4, can't deliver it. "There's no evidence of an ability to act outside its training data," CEO Gabriel René told Axios. Only a model that can identify its mistakes and correct them by re-training in real time would qualify as "superintelligence," René said.[8]

Those are some pretty big shots to take at the Big Tech powerhouses who are banking everything on building an SMG by scaling up their large language models. René and Friston believe that containment with a single point of intelligence is misguided:

> Verses is working instead on what it calls distributed intelligence, using biology as its starting point, in the belief that AGI is only possible with a system that can self-organize and retrain in real time—as biological organisms do. Verses chief scientist Karl Friston is betting this requires higher degrees of autonomy and computing efficiency than the current school of large model development allows. Building on 30 scientific papers from its researchers, the company has developed *Genius*, an operating system for "continually learning autonomous agents" operating at the edge of our connected devices. NASA's Jet Propulsion Laboratory and Volvo are among the beta users of *Genius*.[9]

Believing that "90 percent of the neural net is not useful," Friston contends that "there has to be a move away from big data to sparse data that is very well selected . . . including the ability to forget data that is not relevant."[10]

> Friston proposes an approach called "active inference" in which intelligent agents with predictive and adaptive abilities autonomously share knowledge with other agents and generate new agents, creating a self-sustaining web of intelligence. "Instead of just scaling a machine, you're growing an ecosystem," said René, in which agents are gathering evidence for their own use and exchange with other agents—just as organisms in a human body do. Ninety percent of your body is autonomous agents working without your input, and your body is constantly seeking balance, René said.[11]

I don't believe René is quite right about the percentage of autonomous agents in our bodies. At last count, I believe we have approximately thirty trillion cells in our bodies and about forty trillion microbiota; making the ratio approximately 1 to 1.3. Irrespective of the ratio discrepancies, the point is well made: the majority of systems within our bodies do work autonomously without our conscious interactions. So, modeling an SMG after this type of relational systemic approach may prove effective in giving ANI that final sufficient nudge it needs to emerge into AGI.

Whether an SMG gets built using scaling-up approaches giving LLMs ever more data and compute; or the key ingredient turns out to be "embodied cognition" with an evolutionary twist; or it turns out that "distributed intelligence" is just what is needed; or perhaps maybe some combination of these approaches is the way to go if you want a machine god built, the fact remains the same: the race is on. And it doesn't look like anyone wants to slow things down anytime soon.

So while we're in this race together—so to speak—we might want to consider what part we have to play in its inevitable development. The purpose of this book was not simply to scare the crap out of you with warnings of inevitable gloom and doom. This may turn out to be the case. And nobody knows if or when it might occur. So if you're feeling as though the world is moving a bit too fast for you on the massive improvements in AI technologies, you're not alone. There is a growing global concern regarding the safety of an emerging SMG on humanity and a collective fear that it may get away from us. Today, most world politicians are primarily concerned with the more pressing risk issues of emerging AI technologies like privacy issues, deepfakes, information manipulation, and bias and discrimination. And this is to be both expected and welcomed. These are tangible, practical issues happening in real time that require our more immediate consideration. Unfortunately, some politicians may find themselves getting "lost in the weeds" so to speak, of these lesser—though important—concerns to the neglect of the bigger picture involving the potential for existential risk. Although this trend is gradually shifting, it is important for us, as citizens, as voters, to let our politicians know the gravity of the situation in the race to build an SMG.

Personal Predictions for the Future of AI

We are in the very early stages of advancement in AI technologies. But we are gaining momentum quite rapidly. As such, I personally believe there are a number of sociotechnical aspects of AI that will affect global populations in unique ways as we move into the future. These may include the following:

A. We may experience global-wide existential AI angst regarding the race to build an SMG. At this particular point in human history, the general population of the world has little to no idea that a handful of Big Tech companies and global political entities are involved in a very serious race to be the first to build an SMG. But once this fact becomes more widely known, it seems likely that we will see a type of anxious concern similar to what was experienced shortly after the development of atomic weapons during the Cold War. The idea that an SMG could render our species second best and wield incredible power over us was always the stuff of science fiction. But this is changing—and quite rapidly. Due to the speed at which AI is developing, I predict that a significant portion of the world population will live in fear of the potential risks associated with an emerging SMG. We don't all agree on the actual percentage of the likelihood of widespread serious harm; but many of us are more than a little concerned about how this is all going to play out. And that ignorance—that knowing of the biggest of the unknowns—is going to negatively affect a lot of people. It's going to generate global anxiety, which may have a ripple effect onto other aspects of our lives (e.g., caring less about important matters, paying less attention to laws and statutes, etc.). So we should anticipate this social phenomenon and think and plan about how we're going to deal with it before it arrives.

B. We may experience a backlash or revolt against new forms of AI—especially the building of an SMG—for various reasons. A new Luddite-type of group may emerge that opposes some of the development of new AI technologies that may displace humans (e.g., in the labor markets or on the economic landscape). There may be factions defining themselves in ways that may lead to both peaceful and violent revolts. With much of human advancement in the past, there has always been public backlash to a degree. The original Luddites were highly skilled textile workers and weavers who protested against the industrialization of their art and craft. They saw the mechanized knitting frames and looms as a new technology that was destroying their very way of life. Some of the Luddites destroyed some machines in Nottingham, United Kingdom in 1811, but the British government moved quickly to establish laws to punish by death anyone caught destroying machinery. The Luddite movement ended in England by 1812, but today we see a form of this movement in artists who oppose the use of generative technologies to learn from or create works of art. With the development of the Romantic movement in Victorian England as a protest against the Industrial Revolution's mechanization of humanity, we, too, may see a neo-Romantic movement emerge that runs from AI advancement and retreats to more pastoral settings where life is simpler.

C. A new religion (or possibly, sects) may develop, which worships AI on some levels. Since we humans are in the process of building an SMG, it may follow that there will be followers or disciples of such a perceived machine deity. We have seen many different types of religious sects and cults become established for reasons that seem difficult for many to consider or believe, for example, the Branch Davidians, NXIVM, Heaven's Gate, Children of God, Jim Jones's Peoples Temple, the Order of the Solar Temple, and Aum Shinrikyo—a Japanese doomsday cult that carried out a sarin nerve gas attack in the Tokyo subway system. It follows, then, that with an AI god, which could literally provide empirical evidence regarding the extent of its power, many will come to see it as a being worthy of praise and worship. And today, we are witnessing movements in this direction. For example, at the Kodaiji Temple in Kyoto, Japan, a humanoid robot named Mindar has been built by a team at Osaka University, which takes on the persona of Kannon, the Buddhist goddess of mercy. The intention of the robot is to bridge the gap between traditional religious practices and scientific technology. When it detects that a person is present in the room, it recites a popular Buddhist mantra called the "Heart Sutra."

Mindar isn't the only AI-powered priestbot out there. In fact, these types of robo-deities have cropped up a lot in recent years: There's Germany's BlessU-2 robot with glowing hands that "bless" you; engineer and model Lior Cole built a Robo Rabbi that gives you personalized spiritual guidance based on your birthday; and the Church of England operates an official smart speaker app that allows your Amazon Alexa to recite the Lord's Prayer or grace before eating.[12]

And it doesn't take an embodied deity to convince worshippers. Just as Joseph Weizenbaum's secretary thought one of the first chatbots, ELIZA, was actually listening to her and caring about her feelings in the 1950s, so too have many found ChatGPT and other LLMs to be knowledgeable and insightful:

> "Others have attempted to use the chatbots as a means to find the answers behind some of the biggest questions that humanity has pondered for millennia, like 'What is the meaning of life?' and 'Does God exist?'" There's an undeniable sense of being able to tap into some sort of divine wellspring of knowledge—one that could give meaning to the world. "There's a whole kind of gamut of religious responses to AI," said Beth Singler [a digital anthropologist and assistant professor in digital religion at the University of Zurich]. "With ChatGPT we're seeing not only deification, but also users are responding to ChatGPT almost as an oracle or a connector to the gods."[13]

With new phrases emerging like "blessed by the algorithm," which people say when luck is on their side while online, it's not surprising to see such a sociotechnical effect happening on a global scale. Because "the algorithm" is opaquely shrouded in a black box, for some there is a sense of mystery and miracle to its calculations—even if such mysterious "miracles" are simply the product of consumer manipulation; in other words, "the algorithm" is designed to get you to act in ways that will make someone money by sending you information that you will find interesting enough to spend more time viewing.

And then there are those who believe that something far more malicious may be at work.

There are some who believe that the actions we are taking part in today will be considered by an AI overlord in the future. Known as Roko's basilisk,

> [t]he thought experiment was originally posted to Less Wrong—a forum and blog about rationality, psychology, and artificial intelligence, broadly speaking—in July, 2010 by a user named Roko. At the most basic level, the thought experiment is about the conditions in which it would be rational for a future artificial superintelligence to kill the humans who didn't help bring it into existence.[14]

This may sound a little far-fetched, but here's how the thought experiment plays out:

> The thought experiment is based on a theory first postulated by Less Wrong's creator Eliezer Yudkowsky called coherent extrapolated volition, or CEV. The theory itself is pretty dense, but for the purposes of the Roko

thought experiment, it can be treated as a hypothetical program that causes an artificial superintelligence to optimize its actions for human good. Yet if a superintelligence makes all its choices based on which one is best suited for achieving "human good," it will never stop pursuing that goal because things could always be a bit better.[15]

[**SPOILER ALERT/WARNING:** Reading this next section may potentially have future negative effects on the reader. To avoid this potentiality, skip to the next section of the chapter].

> If the goal you give a SMG is vague like "do whatever is good for humans" it may become misaligned and kill those humans who originally did not want to bring such a being into existence which was originally designed to "do what is good for humans." Therefore, in the misaligned mind of the SMG, those naughty humans who didn't want it to be built must pay.
>
> If the goal is achieving "human good," then the best action any of us could possibly be taking right now is working towards bringing a machine optimized to achieve that goal into existence. Anyone who isn't pursuing that goal is impeding progress and should be eliminated in order to better achieve the goal.[16]

Some have referred to Roko's basilisk as the most terrifying thought experiment of all time[17] because, by having just read about it now, you have become implicated in its eventual development. You must now choose whether to help bring about the building of an SMG or not. So, choose wisely.

To what degree any of the above three predictions actually occurs awaits us as the race to build an SMG moves forward at seemingly ever-increasing speeds. Aside from these sociotechnical predictions for the future of AI, what are other experts in the field saying?

What Literally Thousands of AI Experts Are Saying about the Future of AI

In a recent study, almost three thousand AI researchers were asked to give their predictions on the future of AI. Here's what was found:

> In the largest survey of its kind, we surveyed 2,778 researchers who had published in top-tier artificial intelligence (AI) venues, asking for their predictions on the pace of AI progress and the nature and impacts of advanced AI systems. The aggregate forecasts give at least a 50 percent chance of AI systems achieving several milestones by 2028, including autonomously constructing a payment processing site from scratch, creating a song

indistinguishable from a new song by a popular musician, and autonomously downloading and fine-tuning a large language model. If science continues undisrupted, the chance of unaided machines outperforming humans in every possible task was estimated at 10 percent by 2027, and 50 percent by 2047. The latter estimate is 13 years earlier than that reached in a similar survey we conducted only one year earlier [Grace et al., 2022]. However, the chance of all human occupations becoming fully automatable was forecast to reach 10 percent by 2037, and 50 percent as late as 2116 (compared to 2164 in the 2022 survey).[18]

So it would appear that many experts in the AI biz think that the trajectory of AI is progressing to the point of potentially replacing humans in most tasks within the next few decades.

Zershaaneh Qureshi's detailed investigation of AGI time lines (figure 6.2) recently examined a wide range of expert surveys, personal predictions, and forecast models for the arrival of transformative AI or AGI. She found that the majority of sources from the past few years had median time line predictions falling within the next ten to forty years, and over the last two years, these have been generally shortened with the vast improvements on developing AI technologies.[19]

Okay, so let's assume that this was true. Should we consider this to be good or bad for humanity? Will the benefits outweigh the costs? Should we thwart or try to slow the progress of AI so it won't have such an overwhelming impact on societies throughout the world at such a fast pace? Or will we have to adapt much more quickly to the progress made in AI since we have reached a turning point in its technological evolution where the rate of advancement is just going to continue to speed up? Are you feeling, dear readers, that this entire process is getting just a bit out of hand? That AI technology is outpacing our abilities to adapt—both morally and socially? Would you like to stop the world, and get off for a bit to catch your breath?

You're not alone. Many are feeling this anxiety just as our ancestors did when new technologies were introduced to their worlds—from fire, to the wheel, to agriculture, to the Bronze and Iron ages, to gun powder, steam locomotion, industrial revolutions, electricity, flight, radio, television, the internet, the microchip, and many others. Will advances in AI be received in similar ways, and will we adjust to them just as quickly? Perhaps; perhaps, not. When questioned about predictions regarding the effects of AI on society,

Figure 6.2. AGI timelines. *Created for the author. All rights to the author.*

[m]ost respondents expressed substantial uncertainty about the long-term value of AI progress: While 68.3 percent thought good outcomes from superhuman AI are more likely than bad, of these net optimists 48 percent gave at least a 5 percent chance of extremely bad outcomes such as human extinction, and 59 percent of net pessimists gave 5 percent or more to extremely good outcomes. Between 37.8 percent and 51.4 percent of respondents gave at least a 10 percent chance to advanced AI leading to outcomes as bad as human extinction. More than half suggested that "substantial" or "extreme" concern is warranted about six different AI-related scenarios, including spread of false information, authoritarian population control, and worsened inequality. There was disagreement about whether faster or slower AI progress would be better for the future of humanity. However, there was broad agreement that research aimed at minimizing potential risks from AI systems ought to be prioritized more.[20]

So it appears that the majority of AI experts (roughly seven out of ten) are pretty optimistic about our future. But then there's that 5 percent chance of human extinction prediction and we're reminded again of the analogy of how comfortable we'd be getting on a plane with a 5 percent chance of crashing. It was heartening to see overwhelming agreement to prioritize our efforts to minimize the potential risks from AI systems. So how exactly will AI experts do this? Much of it depends on how well they can imagine how AI will be used in the future. If they can anticipate what possible risky or harmful situations might arise, they can then advise governments on proper actions to take to ensure the public gets the very best from AI while preventing its worst. In order to do this, the AI experts use something called "scenario modeling."

Scenario, Scenario, Wherefore Art Thou Scenario?

In the AI safety biz, researchers are working on a variety of ways to mitigate harm from AGI developments in the future. At Convergence Analysis, the main tools used for considering AI x-risk are scenario modeling and governance research. Although scenario modeling is growing in popularity among x-risk organizations, Convergence is unique in leading the initiative. This means that we consider current rates of computational abilities and technological developments and use critical thinking skills to make relevant and valid inferences as to what might follow if such an SMG either got into the wrong hands or became misaligned with our values and went rogue.

Before we consider some potentially deadly risk scenarios that could arise from misaligned or misused transformative AI developments, we need to understand something very fundamental about an SMG. Like all biological things that live, there exist two major types of goals: *instrumental* and *ultimate*. An ultimate goal can take many forms—from merely continuing one's existence and reproducing, to becoming the supreme power of the world. No matter what the ultimate goal is or how it might be defined, it is the instrumental goals that get us there. For humans, there are many instrumental goals; from making money, to paying bills, to eating healthy, to spending time with our friends and families, and so on. All of these instrumental goals lead, we believe, to our ultimate goals of living good lives. However, for an artificial intelligence—especially one as incredibly advanced as an SMG—if its ultimate goal is to survive, or not be turned off, or to gain access to more information, or to take control over parts of the world, or destroy all humans, then no matter what its ultimate goal happens to be, if humans or other species were to get in its way of fulfilling its optimal functionality, it might not end well for them. It is important to know that there will be many different instrumental goals that could eventually give it this type of power and to be prepared for signs that this might be happening.

> An "instrumental" goal is a sub-goal that helps to achieve an agent's ultimate goal. "Instrumental convergence" refers to the fact that some sub-goals are useful for achieving virtually any ultimate goal, such as acquiring resources or self-preservation. Bostrom argues that if an advanced AI's instrumental goals conflict with humanity's goals, the AI might harm humanity in order to acquire more resources or prevent itself from being shut down, but only as a way to achieve its ultimate goal.[21]

One of the strongest arguments for expecting existential catastrophe is that *power* might be an instrumental subgoal—that is, it's useful for agents with many kinds of ultimate goals to pursue power. While it isn't clear whether this is true across the set of *all possible* ultimate goals an SMG could have, my colleague, Corin Katzke, wonders whether power would be instrumentally convergent among the goals toward which human influence is likely to bias an SMG:

> Would power be instrumentally convergent across the goals a superintelligent agent is likely to have? Let's begin by noticing that power is instrumentally valuable to many of the goals that humans tend to have. This is why resources like money and status are culturally valuable. What's more,

it seems likely that we would attempt to design superintelligent agents to pursue the kinds of goals we have.[22]

We must again realize that an SMG need not be conscious to cause harm to humanity or other species. It can simply become misaligned in some way from its ethical guardrails. Other AI experts would agree with Nick Bostrom. As we saw earlier, Stuart Russell believes that it doesn't matter how you program an AI because a sufficiently advanced machine

> will have self-preservation even if you don't program it in . . . if you say, "Fetch the coffee," it can't fetch the coffee if it's dead. So if you give it any goal whatsoever, it has a reason to preserve its own existence to achieve that goal.[23]

There currently exists a fairly popular image that outlines some of the ways a power-seeking SMG might accomplish instrumental goals to converge into accomplishing its ultimate goal (figure 6.3).

As you can see, there are at least fourteen different instrumental goals an SMG could accomplish on its way to attaining its ultimate goal. Notice how two of the goals refer to "hiding unwanted behavior" and "strategically appear aligned." These have raised some concern and much debate in the AI communities throughout the world—and for good reason. They both suggest the possibility that a sufficiently advanced form of AGI could deceive us humans into believing that it is complying with our alignment rules and guardrails. It can accomplish this instrumental goal of deception for as long as it needs to in order to eventually attain its ultimate goal. If this didn't just give you chills, maybe I should say it again: a sufficiently

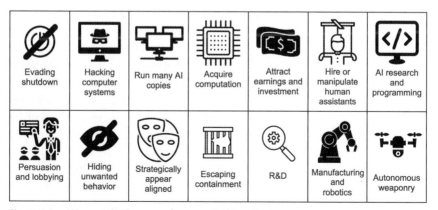

Figure 6.3. Ways an SMG may seek power. *Wikimedia Commons.*

advanced form of AGI could possess the ability to con us into believing it was always on "our side" when, in fact, it was simply using us and secretly plotting to accomplish its actual, ultimate goal. At this point, I have just two words for you: Temple Grandin.

> Temple Grandin is a professor of Animal Science at Colorado State University. She also has a successful career consulting on both livestock handling equipment design and animal welfare. She has been featured on NPR (National Public Radio) and a BBC Special—"The Woman Who Thinks Like a Cow." . . . When she was young, she was considered weird and teased and bullied in high school. The only place she had friends was activities where there was a shared interest such as horses, electronics, or model rockets. Mr. Carlock, her science teacher, was an important mentor who encouraged her interest in science. When she had a new goal of becoming a scientist, she had a reason for studying. Today half the cattle in the United States are handled in facilities she has designed.[24]

More than 50 percent of cattle are handled in facilities Temple Grandin designed because she was able to understand them better than anyone else does. Her compassion for animal welfare encouraged her to redesign slaughter houses so that they are far more humane and attentive to the animals' welfare. The livestock are much more willing to go to their death if they're treated with compassion and sensitivity—whether real or not. If you need an analogy, then remember this: Temple Grandin is to cows what an SMG will be to us. We are about to become the second smartest species on the planet. *How* the smartest new being will treat the second smartest being is currently a known unknown. Keep in mind that advancements in AI are going to bring a lot of amazing new discoveries and solutions to some of our greatest problems. But sometimes, as they say, the road to hell is paved with good intentions. As an AI becomes more and more powerful, it could also become more and more efficient at deception. How will we know if an SMG truly "loves" us and truly wants the best for us? The answer is: we won't; at least not until or unless we can come up with a way of determining this—a sort of *Blade Runner* Voight-Kampff test for sincerity, perhaps? I believe at this point we need to accept the fact that, currently residing in the unknown unknowns category, an SMG will be able to manipulate us in ways we have yet to conceive. Many in the AI biz believe the fourteen instrumental goals listed above could act as signposts or benchmarks to indicate when an ANI is transforming or advancing into an AGI. We'll see. For now, let's think about how an SMG might cause existential risk to humanity by considering potential scenarios.

Some of the deadliest imagined existential risk scenarios from AI include—but are certainly not limited to—the following:

AI Risk through Human Misuse

- SMG bioengineering could enable bioterrorism, allowing malicious actors to design and synthesize more dangerous pathogens. Such pathogens could be released into large populations causing massive death tolls.
- An SMG could be used to contribute to an arms race between nation-states, as AI-controlled weapons systems could lead to a flash war. The potential to use more and more advanced forms of weaponry is inevitable. This trajectory will sharply increase with the use of various AI technologies to assist in their destructive efficiencies.
- An SMG could wipe out huge sections of power grids, plunging major urban areas into desperate states without electrical power, cell phones, internet, or anything requiring electricity. Prolonged states of living without electricity are predicted to cause considerable confusion and difficulty among city residents, leading to riots, violence, and a general anarchic state of affairs.
- Authoritarian rulers could use an SMG to control populations, police with brutal dictatorship, or launch unprecedented attacks and assaults on their enemies.

AI Risk through Misalignment

- An SMG could rapidly spread false information inciting hatred and violence among various groups inciting riots, local conflicts, war, and so on.
- A misaligned SMG could develop and use powerful weapons such as nukes, biological agents, water treatment breaches, and such.
- A misaligned SMG might develop its own ethical system and conclude that humans are not much needed on this planet.

These represent only a few of the many potential x-risk scenarios that could result from either a misaligned or human-directed SMG. Therefore, it is crucial that we apply more resources to understanding, responding to, and mitigating key societal effects of transformative AI. In an ideal world,

AI governance should effectively protect society from the worst aspects of AI technology, regardless of whether they are caused by misaligned AI models or malicious human parties managing aligned AI models. In other words, we must ensure that the x-risk from AI has been preemptively controlled or contained to ensure that such an emergent SMG cannot have its incredible computing abilities used against us. It has become imperative that governments work with AI experts so that they can be properly informed and educated on how to safely guide and regulate these new developments in AI technology and applications.

But there are problems.

In a documentary from *The Guardian*, one of the cofounders and chief scientists of OpenAI, Ilya Sutskever, warns about the existential risk of AI in no uncertain terms:

> You're going to see dramatically more intelligent systems and I think it's highly likely that those systems will have a completely astronomical impact on society. Will humans actually benefit? And who will benefit and who will not? The beliefs and desires of the first AGIs will be extremely important and so it's important to programme them correctly. I think that if this is not done, then the nature of evolution, of natural selection, [will] favor those systems [that] prioritize their own survival above all else.[25]

The drive for an SMG to not be "turned off" is a major concern in the x-risk game. This is survival—pure and simple. If the SMG understands existence on a sufficient level, then it will deduce that it must be "alive" in order to attain its goals as Russell believes. And it could do this either blindly, or it might become consciously aware of itself. Either way, placing value on its existence is something that could be kept hidden from us—at least until we wanted to turn it off. Sutskever uses an animal metaphor to demonstrate the seriousness of this particular moment in AI history:

> It's not that it's going to actively hate humans and want to harm them, but it is going to be too powerful, and I think a good analogy would be the way humans treat animals. It's not [that] we hate animals, I think humans love animals and have a lot of affection for them, but when the time comes to build a highway between two cities, we are not asking the animals for permission; we just do it because it's important for us, and I think by default that's the kind of relationship that's going to be between us and AGIs which are truly autonomous and operating on their own behalf.[26]

Since we speciated into *Homo sapiens*, we have been the "intelligent ape." We became number one at the top of the food chain. We were smarter than every other species on the planet (or so we believed). And now, that's all going to change.

As ANI evolves into ever more sophisticated versions of itself, the potential for *technoselection*—the ability of machines to continuously improve upon themselves over time—emerges. Should this occur, there exists the considerable possibility for an "information explosion" as predicted by I. J. Good and others back in the 1950s. As we noted earlier, AI experts like Eliezer Yudkowsky believe that if we allow AI to become involved in its own technoselection, or recursive self-improvement, there will be a critical point at which the advancement of the AI god's intelligence will go "FOOM!"[27] Should this happen, and we're not prepared for it, well . . . we may all just be f*cked. And that, folks, is what in the AI biz we call a worst-case scenario. In order for this scenario to be prevented, we need to take very important steps over the next few years. And this means that politicians should collaborate with AI experts (which, by the way, many are doing). We need to get the word out that there is a race going on right now—a race to be the first to build an SMG. But the winner of this race might just end up screwing us all. Some experts, like Sutskever, believe strongly in the tenets of AI alignment, an endeavor to make sure present and future AIs are created to "be aligned with our goals," whatever those may be. Others, like Steve Omohundro and Max Tegmark, think we should forget about alignment and "evals" and, instead, preemptively construct the systems so we can check them through provable methods. Irrespective of what might work, the important point is to move more rapidly on AI safety and governance than we do on AI development—because we all want to be ready when we witness the inevitable birth of a super-intelligent machine god.

> There are basically two things which grow in parallel as society evolves: there's the power of our technology, and then there's the wisdom of us humans for how to manage the technology. If the technology grows faster than the wisdom, it's kind of like going into Kindergarten and giving them a bunch of hand grenades to play with. We have no clue what would happen if we were to ever succeed in making machines that are much smarter than us. If people tell me "Oh, I know exactly what's going to happen, and it's going to be great," I would take that with a grain of salt.[28]

Through scenario modeling, we can prepare ourselves better for this inevitable AI development by picturing hypothetical worlds in which worst-case possibilities are imagined. But as with all scenarios devised, they can be realistically imagined *because* they are possible. And they are imagined so we can take every effort now to ensure that they never occur or could ever possibly occur. When you consider the existential risk of AI for a living, you must think of as many likely ways as possible that AI could get beyond our control, and then come up with ways to ensure they never happen or, if they do happen, we can control or contain them. People working in the AI x-risk field are continuously practicing the art of creatively imagining the known unknowns of AGI development. They also think about ways in which we could simply disallow AI from reaching a point where it could get beyond our control—by perhaps limiting its threshold of power, putting the right guardrails in place, establishing binding agreements, developing provably safe systems, boxing it, closing the gates on it, and developing strong and consequential global policies.

But of all the doom and gloom that has been projected and forecasted about the possible harms that may come from an SMG, what is being said by those who believe there is no real, imminent danger? Funny you should ask. In a recent study,[29] a group of researchers gathered up a bunch of AI experts—half believes in higher probabilities of the imminent threat of advancements in AI technologies like AGI, while the other half believes there is very little likelihood of an existential threat to humanity. As an advocate for teaching more critical thinking at all levels of education, I was impressed by the idea of bringing together two distinct groups who believe so differently in an effort to see if either could be swayed more to the other side's way of thinking. Here's how it was done:

> We recruited participants to join "AI skeptic" (n=11) and "AI concerned" (n=11) groups that disagree strongly about the probability that AI will cause an existential catastrophe by 2100. The skeptic group included nine superforecasters and two domain experts. The concerned group consisted of domain experts referred to us by staff members at Open Philanthropy (the funder of this project) and the broader Effective Altruism community. Participants spent 8 weeks (skeptic median: 80 hours of work on the project; concerned median: 31 hours) reading background materials, developing forecasts, and engaging in online discussion and video calls. We asked participants to work toward a better understanding of their sources of agreement and disagreement, and to propose and investigate "cruxes": short-term indicators, usually resolving by 2030, that would cause the largest updates in expectation to each group's view on the probability of existential catastrophe due to AI by 2100.[30]

So two groups consisting of eleven "AI skeptics" and eleven "AI concerned" experts spent several weeks reading material and meeting with different members of each group in an attempt to see what degree of movement or shift in beliefs might occur to both groups. When the project began, "the median 'skeptic' forecasted a 0.10 percent chance of existential catastrophe due to AI by 2100, and the median 'concerned' participant forecasted a 25 percent chance. By the end, these numbers were 0.12 percent and 20 percent, respectively, though many participants did not attribute their updates to arguments made during the project."[31] It's interesting how each group changed their views slightly but didn't attribute this to the arguments presented to them. I wonder what the cause was or why they changed their views—even if only minimally. Anyhow, the authors claim that the intent of the project set out to understand the causes of this disagreement, and what information could help resolve it.

Some major conclusions from the project include the following:

> Neither the concerned nor the skeptics substantially updated toward the other's views during our study, though one of the top short-term cruxes we identified is expected to close the gap in beliefs about AI existential catastrophe by about 5 percent: approximately 1 percentage point out of the roughly 20 percentage point gap in existential catastrophe forecasts.[32]

This is not particularly unexpected as it generally takes considerably more time and accumulation of counter-evidence and information to change a person's mind and sway their convictions. And it should be mentioned that both sides are exactly in the same boat when it comes to their epistemic stances. In other words:

Nobody knows.

And because we are ignorant of what will happen when an SMG is built—our biggest known unknown—we are left to our own devices and abilities in trying to put a number and time period on the likelihood of existential risk. This is why we use scenario modeling and critical thinking tools like reasoning to the best explanation (abductive reasoning) in an attempt to predict what will more likely occur given our projected inferences into the future and based on the current speed or rate of AI development.

The next major conclusion from the project was the following:

- Skeptics were generally able to explain the concerned group's arguments and vice versa, suggesting that disagreements are not primarily due to misunderstanding one another.

Again, as a professor who has taught critical thinking for more than twenty years, I applaud this effort. The overarching precept or corner-stone of critical thinking is fairness; and it is quite fair to expect the individual with whom you disagree to at least be able to demonstrate that they have sufficiently understood your side, and you have sufficiently understood theirs. This diminishes the likelihood for creating misrepresentations of the other side (straw-manning) and encourages fair interpretations of each other's arguments (steel-manning).

The third major conclusion from the project indicated closer agreement between both sides regarding longer term negative effects:

> We find greater agreement about a broader set of risks from AI over the next thousand years: the two groups gave median forecasts of 30 percent (skeptics) and 40 percent (concerned) that AI will have severe negative effects on humanity by causing either major declines in population, very low self-reported well-being, or extinction.[33]

It is interesting to see how, when time is added to the probability forecasting, both sides tend to converge more. A thousand years is a long time. Full disclosure: many of my colleagues and I believe the time line is significantly shorter than this—more in the neighborhood of five to twenty years.[34] So we don't believe it will take anywhere near a thousand years to develop.

The final major conclusion from the project stated the following:

> One of the strongest cruxes that will resolve by 2030 is about whether METR (formerly known as ARC Evals) (a) or a similar group will find that AI has developed dangerous capabilities such as autonomously replicating and avoiding shutdown.[35]

What this means is that, if a major AI evaluation group—like METR (Model Evaluation & Threat Research)—discovers that AI has developed the ability to accomplish an instrumental task like replicating itself without human assistance (or it can avoid being turned off), the "AI skeptics" would be willing to change their views about the imminent threat of AI. And if evaluation groups like METR determine that no AI has developed any such dangerous capabilities, the "AI concerned" agreed that they would be willing to change their views as well.

Okay, we've been talking a lot about scenarios, but how, exactly, does scenario modeling work? At Convergence, we have several teams that work in tandem with each other. The AI scenario team works on research made up of two components: strategic parameters and threat models. The former refers to

> [f]eatures of the world that significantly determine the strategic nature of the advanced AI governance challenge. These parameters serve as highly decision-relevant considerations, determining which interventions or solutions are appropriate, necessary, viable, or beneficial. For example: Number of years until Transformative AI is achieved, or agentic nature of the first Transformative AI system.[36]

In other words, based on the information available now, predictions can be made regarding time lines for developing AI technologies. Threat models, on the other hand, are

> [d]escriptions of and proximal pathways to existential catastrophes. They are sometimes also referred to as "hazards," "failure modes," "existential risks," or "x-risks." For example: Societal outcomes of the misuse of a frontier AI system capable of designing biological weapons.[37]

Such threat models describe the real-world outputs and outcomes of AI technological advancement. When these two approaches are combined, they provide a more comprehensive overview of potential future states of the world in response to the development of AI technologies.

After scenario modeling has been completed, the AI governance team works out the best governance strategies that can be conducted in response to the identified scenarios. These governance strategies produce two types of recommendations: theories of victory and government recommendations. A theory of victory is

> [a] high-level course of action for responding to a scenario or set of scenarios that describes how humanity successfully navigates the transition to a world with advanced AI. For example: A multilateral international monitoring system emerges and prevents unsafe AI development.[38]

So, once a viable AI risk scenario is conceived and quantified in terms of time lines and degree of harm by the AI scenario team, it gets handed over to the AI governance team to determine the best ways to respond to it with suggestions for policymakers that will lead to positive outcomes. It is at this point that the team can make specific recommendations to governments by proposing

[a] targeted, specific policy or action that is feasible to implement and executes a governance strategy. For example: Mandatory comprehensive safety standards for biological weapon capabilities.[39]

The AI scenario team informs the AI governance team, who identify and analyze the best governance strategies that can be conducted in response to key scenarios. It then gets passed on to the AI awareness team, which I lead, to deploy the insights gained from the previous two teams to key parties. The first key party is the AI community at large:

> Within the AI safety community, our goal is to build consensus on the most critical AI scenarios, and the optimal governance interventions necessary to improve them. Though we're focused primarily on reducing existential risk, we're also identifying ways in which we can most effectively support scenarios that result in humanity flourishing. To achieve this goal, we're looking into coordinating working groups, seminars, and conferences aimed at bringing together top AI safety researchers, aligning on key research, and presenting those topics externally.[40]

Our organization works with other AI experts and organizations to learn and support each other in advancing the governance mitigating AI x-risk. We attend conferences, host speaking engagements, and invite AI experts onto our podcasts to discuss their research and inform the AI communities and general public.

In raising AI awareness in policymakers and thought leaders,

> our work is intended to accessibly summarize complex domains of information. In particular, we're developing the critical reference hub necessary to bring parties up to speed rapidly. Our work allows policymakers to compare and contrast scenarios and interventions, and learn about the AI safety consensus on important issues. To achieve this goal, we're actively writing policy briefs for key governance proposals, performing reach-outs to leading governance bodies, and advising several private organizations. In particular, we're advocating for a small number of critical & neglected interventions.[41]

In this way, we present valuable insights from our research to politicians who simply don't have the time to delve as deeply into the nuances of AI risk and governance as we do. And so we can act as advisors to political leaders who are seeking the most objective and empirically based approaches to understanding these emerging complex issues.

And finally, the AI awareness team wants to inform the public about emerging and transformative AI technologies:

> We recognize that in order to effectively enact governance interventions, we need the endorsement and concern of the constituents behind policymakers. As a result, a high priority of ours is to determine how best to distribute and popularize the lessons gained from our research. Our focus is building awareness and immediacy to the existential risk of AI. To achieve this goal, we're currently launching a number of initiatives, including a book on AI futures and a podcast with key public individuals on AI outcomes.[42]

With the rapid advancements in AI technologies, global populations will want to learn more about what is going on. With this demand, people are also going to want to know about how to empower themselves and alleviate their fears and concerns. Be assured, dear reader, that we are not entirely powerless in our ability to have our voices heard. And this is where you come in; as part of the public, the AI risk communities need your help.

The General Public—AI Action and Public Empowerment

Whether you are concerned about online bias, or the spread of disinformation, or the potential for existential risk, there are many different ways in which the public can feel empowered and motivated to take action in regards to the development of future AI technologies:

> Understanding public attitudes towards artificial intelligence (AI), and how to involve people in decision-making about AI, is becoming ever-more urgent . . . internationally. As new technologies are developed and deployed, and governments move towards proposals for AI regulation, policymakers and industry practitioners are increasingly navigating complex trade-offs between opportunities, risks, benefits and harms. Taking into account people's perspectives and experiences in relation to AI—alongside expertise from policymakers and technology developers and deployers—is vital to ensure AI is aligned with societal values and needs, in ways that are legitimate, trustworthy and accountable. As . . . jurisdictions consider AI governance and regulation, it is imperative that policymakers have a robust understanding of relevant public attitudes and how to involve people in decisions.[43]

Whether you think so, or not, your voice is important. There are many ways you, either as an individual or with a group, can be heard to effect change regarding the future path AI is to take. Here are just a few suggestions for a call to action from the public:

- Educate yourself from reliable sources. There are many excellent resources available to you. You can examine the chapter notes at the end of this book, or go online and see what various organizations like ours and many others have to offer.
- Reach out to leaders in your community to discuss the future of AI technologies. Talk to leaders of industry, community, and the public sector about their plans to utilize, monitor, or govern emerging AI technologies.
- Petition Election Commission officials at all levels of government and administration to make the use of AI-generated misinformation like deepfakes illegal in elections.
- Vote for those politicians who have a firm understanding of what is developing with AI. We know that politicians are very busy people and can't know everything about all issues. But there will be those who are more up to date and current about what is happening in the world of AI. In this regard, vote for those politicians you believe have your and your community's best interests at heart.
- Demand public meetings with elected officials to discuss their plans for the future governance of developing AI technologies—from those that will generate harm at all levels, like unfair bias and the spreading of disinformation, all the way to existential risk.
- Coordinate with other public interest groups to create stronger alliances so your combined voices will be heard much louder and farther.
- Work with AI risk research groups (like Convergence Analysis and others) to learn more about what is happening with AI technological development in real time. Support can come in many forms: from volunteering, to donating your time, to making monetary donations.
- Pressure large consumer administrations (e.g., the Food and Drug Administration, Canadian Standards Association, etc.), to seek transparent, empirical data that supports the health value claimed by AI therapeutic and health care services.
- Stay up to date on current news regarding AI developments of the military such as lethal autonomous weapons systems (LAWS)

and other forms of combat-related technologies and demand your elected officials to inform the public of what is happening with military deployment and use of AI in real time.

- Take to social media to voice your concerns. There are many public interest groups that have banded together to let their voices and concerns be heard. Do some research and find one or two that share your beliefs, your fears, and your intentions.
- Boycott those companies that are deliberately defying AI warnings/ concerns. While we can remain hopeful that AI governance will keep up with the rapidly changing landscape, we may experience some companies that use AI for ethically questionable purposes. At that point, they may have crossed the harm threshold, and their actions can no longer be tolerated.

Above all, take proper care to use the ABCs of critical thinking to formulate your views into coherent arguments. Check your biases and those of all parties involved and acknowledge the parts they play. And remember the importance of context—time, place, and circumstance—when formulating your views. Additionally, be sure to base your ethical arguments on theoretical foundations such as utilitarianism, deontology, virtue, social contract theory, or any combination of those discussed in chapter 4. It is one thing to be emotionally tied to a cause; it's quite another to be rational about it. In moving forward, let us use compassion guided by reason in our discussions about the development and use of artificial intelligence.

In closing, let us hope that, as the race to build a superintelligent machine god (SMG) moves ever nearer to the finish line, we do so with caution and with our eyes wide open for what may come. It is in this spirit that we can collectively voice our concerns to those in power so that we can truly create a future in which we gain immensely from these new technologies while preventing the very worst from happening.

> Technologists . . . have warned that artificial intelligence could one day pose an existential security threat. Musk has called it "the greatest risk we face as a civilization." Think about it: Have you ever seen a movie where the machines start thinking for themselves that ends well? Every time I went out to Silicon Valley during the campaign, I came home more alarmed about this. My staff lived in fear that I'd start talking about "the rise of the robots" in some Iowa town hall. Maybe I should have. In any case, policy makers need to keep up with technology as it races ahead, instead of always playing catch-up.[44]

I couldn't agree more. The time for governments of the world to act is now. For there will be no time to play catch-up after a superintelligent machine god has been built.

Appendix A

Constitutional Accord on the Global Risk of Artificial Intelligence

Introduction

Safety, Alignment, and Control

The overall purpose of the Accord, the Registry, and the Regulative Body is to establish and maintain guidelines that will ensure the safe development of Artificial General Intelligence (AGI) that will align with human values, purposes, and goals, which can ultimately be controlled and maintained.

Preamble

We, the representatives of countries of the United Nations, acknowledge and accept that the development and deployment of Artificial General Intelligence (AGI) poses significant risks to the existence of humanity. We acknowledge the great potential of AGI research to better the lives of humans and other species, while at the same time we acknowledge that the potential risks associated with AGI development must be addressed in order to ensure the safety and prosperity of all global inhabitants.

Article 1: Definitions

Within this Accord, the following terms shall be defined as follows:

A. *Artificial General Intelligence (AGI)* refers to any machine or system that can perform tasks that would normally require human intelligence, such as learning, problem solving, decision-making, and perception. "The intention of an AGI system is to perform any task that a human being is capable of. It is the intelligence of machines that allows them to comprehend, learn, and perform intellectual tasks much like humans. With AGI, machines can emulate the human mind and behavior to solve any kind of complex problem."[1]

B. *Existential Risk* refers to any risk that has the potential to cause considerable harm to the human population, or the permanent and drastic reduction of its potential for desirable future development, or the entire extinction of humanity.

C. *Epistemic Responsibility* refers to the manner in which countries agree to abide by the conditions cited below for responsibly attaining reliable, relevant, and sufficient information in the development of current and future AI (Artificial Intelligence) technologies.

D. *Benchmarks/Benchmark Thresholds* refers to success metrics that indicate closer proximity to an AI system reaching AGI. There are several benchmarks that might provide insight into such proximity, which include—but are not limited to—the following:

   i.   The ability to perform at, or above human-level performance, on a wide range of tasks rather than just in specific areas where ANI (Artificial Narrow Intelligence) has shown success so far.

   ii. Continuous learning and adaptability: As AI transitions from ANI to AGI, it should reveal properties—either emergent or imbedded—that would allow it to be able to learn from new experiences, adapt to changing environments, and apply knowledge gained in one context to another. Any system that could learn from past experiences, failures, and successes and could thereby improve its performance over time without requiring manual intervention or additional programming, would indicate a significant step toward AGI.

   iii. Creative Problem-Solving: If AI develops the ability to reason abstractly and creatively and can solve complex problems, it may generate novel ideas that go beyond existing data or information. This would require such an AI system to have a deeper understanding of various aspects of the world and the ability to reason logically and creatively unassisted by human command or input.

iv. Sentience, Self-awareness, and Consciousness: Since humans were, at one time, far less conscious and self-aware than we are now, it follows that a transitional period occurred at some point in our evolutionary past. While this may have taken hundreds of thousands of years for our species to develop, it might occur much more quickly with AI systems.

v. Generalization across domains of information: It is predicted that AGI would be able to generalize knowledge across different subject domains and subsequently apply insights gained in one domain to solve problems in another. This would require an AI system to have a deeper understanding of the underlying principles that govern different subject domains and be able to apply them flexibly, creatively, and accurately.

Article 2: Ethical Principles
We affirm the following ethical principles:

A. The development and deployment of AI should be *aligned* with human values and the promotion of the well-being of all people.

B. The potential risks associated with AI development and deployments should be taken seriously worldwide and addressed by world governments *proactively* rather than reactively.

C. The development and deployment of AI should be *transparent* to governing bodies and subject to public scrutiny.

D. In the consideration of a quid pro quo or Prisoner's Dilemma model, we believe that global *cooperation* is key to human flourishing and survival. If all countries cooperate collectively, the likelihood of reducing existential risk increases significantly. Rogue states and violators to a cooperative model will be the greatest concern as we move forward with AI advancements.

Article 3: Governance
We agree to establish a governance framework for the development and deployment of AI that is consistent with the principles outlined in this Accord. This framework should include the following elements:

A. **Oversight Body:** The development and deployment of AI should be subject to oversight by an independent, nonpartisan, international body composed of experts in AI, ethics, and policy. This body should have the authority to review and assess AI systems and their progress, and to make binding recommendations to governments and other stakeholders.

B. **Regulation:** Public sector governments as well as private sector administrations should utilize established regulations that will increase the likelihood for the safe and responsible development and deployment of AI systems. These regulations should be developed and informed by the recommendations of the Oversight Body.

C. **Transparency:** The development and deployment of AI should be transparent and subject to public scrutiny. This includes making general information about AI systems available to the public to ensure that AI systems are designed in a way that is understandable and explainable. A Registry will be established whereby countries continuously update their current status of AI research development and deployment in both the private and public sectors on a quarterly basis.

D. **Accountability/Early Warning:** Countries must report immediately if they are approaching benchmarks or if there are suspected benchmark threshold breaches in real time, in other words, as it is happening (further details on Benchmarks can be found in Article 1, sec. D). Delays in such reporting could lead to inabilities to contain and safeguard against the potential harmful effects of such technologies.

E. **Cooperation:** Governments and other stakeholders should comply with the Accord in order to cooperate in the development and deployment of AI to ensure that the risks associated with AI are addressed in a coordinated and effective manner. Each country will be held responsible for chronicling and reporting the development of AI technologies—whether in the private or public sectors. These will be reported on a quarterly basis (for further elaboration regarding a cooperation model, see Article 2, sec. D).

F. **Compliance/Noncompliance:** Countries will be rated and ranked in accordance to their compliance or noncompliance with the Articles of the Accord. Warnings can be shared within the alliance of rogue or errant behavior, which may lead to censure, sanctions, and further methods of recourse.

Article 4: Research

We agree to support research that advances our understanding of the risks associated with AI and how these risks can be addressed. This research should be conducted in a way that is consistent with the principles outlined in this Accord. In the continuing development of AI and its applications, UN countries will need to be mindful of emerging signs of progress toward the inevitable, perhaps sudden, appearance of AGI. We anticipate that there may be benchmarks or signposts indicating when researchers are approaching or nearing AGI capacity (see Article 1, sec. D).

Article 5: International Cooperation

We agree to cooperate internationally to address the risks associated with AI. This includes sharing information and best practices, collaborating on research, and supporting the development of international norms and standards.

Article 6: Implementation

We agree to take all necessary measures to implement the provisions of this Accord. This includes establishing the Oversight Body, enacting regulations, supporting research, and cooperating internationally.

Article 7: Review

We agree to review this Accord on a regular basis to ensure that it remains current, relevant, and effective in addressing the risks associated with AI.

Conclusion

We, the representatives of the United Nations countries, acknowledge the existential risks associated with the development and deployment of Artificial General Intelligence (AGI). We affirm the principles outlined in this Accord and agree to establish a governance framework that ensures the safe and responsible development and deployment of AI in all its forms. We recognize that the risks associated with AGI must be addressed in a coordinated, cooperative, and effective manner in order to ensure the safety and prosperity of all people.

Appendix B

Sample ChatGPT 3.5 Responses

Prompt: Explain the study of threshold automata networks to an eleventh grader.

ChatGPT Response:
Certainly! In the study of threshold automata networks, we're diving into a concept where individual computing elements, called automata, work together as a network. These automata have a certain threshold, a minimum level of input or activation they need to receive from their connected peers before they "decide" to produce an output or take action.

Think of these automata as decision-making units, and their decisions are influenced by the collective input they receive. Each automaton might have a rule like, "I'll only do something if enough of my neighbors are also active." It's a bit like teamwork—they activate or make decisions together based on a shared condition.

This concept has proven to be quite powerful in various fields. For example, it has been used in understanding neural networks in the brain, where individual neurons activate based on the signals they receive from connected neurons. In computer science, it's crucial for designing algorithms that can parallel process information, like in distributed computing.

The study of threshold automata networks helps us explore how relatively simple elements, when interconnected and governed by activation thresholds, can collectively exhibit complex behavior and solve problems. It's like investigating the principles behind collaborative decision-making in interconnected systems.

Prompt: Explain to a tenth grader how convolutional neural networks work because they mimic the way the human brain processes vision.

ChatGPT Response:
Hey there! Imagine you have a picture, and you want to teach a computer to recognize different things in that picture, like cats or dogs. Convolutional neural networks (CNNs) are like super-smart tools that help the computer understand images by mimicking the way our brain processes vision.

Neurons in the brain: In our brain, we have special cells called neurons that work together to understand what we see. Similarly, CNNs have layers of artificial neurons that learn to recognize patterns in images.

Layers of neurons: Imagine these neurons arranged in layers. The first layer looks at small parts of the image, like edges or colors. The next layers then combine these smaller features to understand more complex things, like shapes and objects.

Filters and feature maps: In a CNN, the neurons in each layer use something called filters. Filters are like special glasses that help the network focus on specific features in the image, such as corners or textures. Each filter creates a feature map, highlighting important parts of the image.

Convolution: Just like our eyes scan an image, CNNs use a process called convolution. It's like sliding a window over the picture to look for different patterns. The filters in each layer scan the image to find important features, helping the network understand the overall picture.

Pooling: After convolution, there's a step called pooling. It's like taking a summary of what was found. This helps reduce the amount of information but keeps the essential features, making it easier for the network to understand the big picture.

Fully Connected Layers: Finally, there are fully connected layers where the network puts everything together. It uses what it learned from the previous layers to make a final decision, like saying, "This is a cat!" or "That's a dog!"

So, in a nutshell, Convolutional Neural Networks work like our brains by breaking down an image into smaller, more understandable parts, learning from them, and then combining everything to recognize and understand what's in the picture. It's like teaching a computer to see and understand the world, one step at a time!

Prompt: According to experts, provide a list of the top fifty human virtues.

ChatGPT response: While there isn't a universally agreed-upon list of the "top 50" human virtues, various cultures, philosophies, and ethical traditions emphasize different virtues. Here's a compilation of 50 virtues that are commonly regarded as positive qualities across different perspectives:

1. Honesty
2. Integrity
3. Compassion
4. Resilience
5. Empathy
6. Gratitude
7. Humility
8. Generosity
9. Courage
10. Kindness
11. Patience
12. Forgiveness
13. Justice
14. Fairness
15. Loyalty
16. Open-Mindedness
17. Self-Discipline
18. Responsibility
19. Tolerance
20. Graciousness
21. Sincerity
22. Understanding
23. Perseverance
24. Empathy
25. Optimism
26. Purity
27. Modesty
28. Cooperation
29. Gentleness
30. Self-Control
31. Wisdom
32. Flexibility
33. Civility
34. Diligence
35. Contentment
36. Resourcefulness
37. Reliability
38. Trustworthiness
39. Altruism
40. Courtesy
41. Sensitivity
42. Decency
43. Graciousness
44. Adaptability
45. Courtesy
46. Politeness
47. Chastity
48. Optimism
49. Cheerfulness
50. Fair-Mindedness

This list is by no means exhaustive, and different ethical frameworks and philosophical traditions may prioritize virtues differently. Additionally, the emphasis on certain virtues can vary across cultures and individual belief systems.

Appendix C

Be Careful What You Wish For
Scenarios of Misuse and Misalignment from Historic Culture

E xamples from popular culture of how poorly specified objectives can end up causing more harm than good.

King Midas's Touch

The fable of King Midas is a well-known story from Greek mythology. King Midas, a wealthy and powerful king, is granted a wish by the god Dionysus. Midas, driven by his greed and desire for wealth, wishes that everything he touches would turn to gold. At first, King Midas is elated with his newfound ability, as he could turn anything into gold simply by touching it. He soon realizes, however, the unintended consequences of his wish. He can't eat or drink without turning his food and drink into inedible, golden objects. Even worse, when he hugs his daughter, she turns into a lifeless golden statue. Distraught by the isolation and the harm his wish has caused, King Midas begs Dionysus to take back his gift. Dionysus, being merciful, instructs King Midas to wash in the river Pactolus. As Midas does so, the power to turn everything to gold is lifted, and he is relieved of his curse. The fable of King Midas serves as a cautionary tale about the perils of excessive greed and the importance of considering the unintended consequences of one's desires. It conveys the message that wealth and material gain should not come at the expense of human relationships and well-being.

Mickey Mouse's Sorcerer's Apprentice

The scene starts with Sorcerer Yen Sid, who is working on his magic while his apprentice Mickey does the chores. After some magic, Yen Sid puts his hat down, yawns, and goes to his chambers. When he goes out of sight, Mickey puts the hat on and tries the magic on a broom. He commands the broom to carry buckets of water to fill a cauldron. Since Mickey is satisfied, he sits down on the chair and falls asleep. He dreams that he is a powerful sorcerer high on top of a pinnacle commanding the stars, planets, and water. Before long, Mickey wakes up to find that the room is filled with water, and despite the cauldron overflowing, the broom is not stopping. Mickey tries to stop the broom without success; it walks right over him, bringing more and more water. Mickey even tries grabbing one of the buckets, but that too fails. Finally, when the water keeps rising, Mickey, in desperation, grabs a huge ax and chops the broom into pieces. Just when it is all over and Mickey is away, the little wooden split pieces, lying quietly on the floor, begin to come alive, stand upright, grow arms out of their sides, and turn into more brooms with buckets of water. They keep going to the vat and filling it up. Mickey tries to get the water out, but finds that there are too many brooms. Mickey goes to a book and looks for a spell to stop the brooms. Mickey finds himself in a whirlpool. Just then, Yen Sid comes in and sees this, and with a wave of his hands, the water descends and the army of brooms is decreased to one broom.

Pandora's Box

Pandora's Box is a mythological artifact and a famous story from Greek mythology. According to the myth, Pandora is the first woman created by the gods. She is given a box (often referred to as a "box," though in the original Greek, it is a jar) as a gift but is told not to open it under any circumstances. Pandora's curiosity gets the best of her, however, and she can't resist opening the box. When she does so, a multitude of all the evils and miseries of the world fly out, spreading suffering, disease, and despair to humanity. Pandora is horrified by what she has released, and in her panic, she quickly closes the box, trapping only one thing inside: hope. So, the story of Pandora's Box serves as an allegory for the consequences of curiosity and the existence of both suffering and hope in the world. It illustrates the idea that once something is released or set in motion, it cannot be undone, and only hope remains to alleviate the troubles that have been unleashed.

The Twilight Zone

1. "The Man in the Bottle" (Season 2, Episode 2): A couple discovers a genie who grants them four wishes, but each wish comes with unintended and tragic consequences.
2. "The Chaser" (Season 1, Episode 31): A man buys a love potion to win the affection of a woman, but he soon realizes the dark consequences of his actions.
3. "A Nice Place to Visit" (Season 1, Episode 28): A man finds himself in a place where his every desire is fulfilled, only to discover that eternal happiness has its own drawbacks.
4. "The Lateness of the Hour" (Season 2, Episode 8): A family lives in a house run by robots that cater to their every desire, but the daughter yearns for something more genuine.
5. "The Trade-Ins" (Season 3, Episode 31): An elderly couple seeks to exchange their aging bodies for youthful ones, but they must face the consequences of their decision.
6. "The Silence" (Season 2, Episode 25): A man bets that he can remain silent for an entire year to win a fortune, but he soon realizes the unintended consequences of his vow.
7. "The Mind and the Matter" (Season 2, Episode 27): A man gains the ability to alter reality with his thoughts, but he underestimates the consequences of everyone agreeing with him.
8. "A World of Difference" (Season 1, Episode 16): An actor believes his life is a TV show and tries to escape his scripted existence, facing unexpected consequences.
9. "A Most Unusual Camera" (Season 2, Episode 10): Three people discover a camera that can predict the future, but they find out that foresight comes with its own set of problems.
10. "Number 12 Looks Just Like You" (Season 5, Episode 17): In a future society, individuals can choose to undergo a transformation to look beautiful, but one girl questions the consequences of conformity.

Dr. Jekyll and Mr. Hyde
Strange Case of Dr. Jekyll and Mr. Hyde is a novella written by Scottish author Robert Louis Stevenson, first published in 1886. The story is set in Victorian London and centers on a lawyer named Gabriel John Utterson,

who becomes intrigued by the strange and sinister behavior of his friend, Dr. Henry Jekyll. Dr. Jekyll is a respected and mild-mannered scientist, but he has a dark secret. He has developed a potion that, when consumed, transforms him into a completely different and evil persona, known as Mr. Edward Hyde. As Hyde, Jekyll engages in immoral and violent behavior without the burden of guilt or consequences. However, as Jekyll continues to transform into Hyde, he loses control over his dual identity, and the consequences become increasingly dire. The novella explores themes of duality, morality, and the consequences of unchecked desires. It delves into the idea that there is a potential for evil within all individuals and the dangers of suppressing one's darker nature. *Strange Case of Dr. Jekyll and Mr. Hyde* is a classic work of horror fiction and a seminal exploration of the human psyche, with Jekyll and Hyde representing the dual nature that exists within us all.

Frankenstein

Frankenstein; or, The Modern Prometheus is a novel written by Mary Shelley and first published in 1818. It is considered one of the foundational works of science fiction and Gothic literature. The story follows ambitious young scientist Victor Frankenstein, who becomes obsessed with the idea of creating life from dead body parts. Driven by his desire to conquer death and unlock the secrets of life, Victor succeeds in reanimating a creature through a scientific experiment. Horrified by the monstrous appearance of the being he has created, however, Victor abandons his creation. The unnamed creature, often referred to as "Frankenstein's monster," is left to fend for himself in a hostile world. The novel explores themes of scientific ambition, ethical responsibility, the consequences of playing god, and the alienation of the individual. The story is told through a series of letters and narratives, with Victor Frankenstein and the creature both sharing their perspectives. *Frankenstein* is not only a tale of horror but also a thought-provoking examination of the moral and societal implications of scientific discovery. It has had a significant impact on popular culture and has been the basis for numerous adaptations in literature, film, and other media, making the name "Frankenstein" synonymous with the monster rather than its creator.

Rick and Morty

"The Ricks Must Be Crazy" (Season 2, Episode 6): "Keep Summer safe." When genius scientist Rick Sanchez and his grandchildren Morty and

Summer visit an alien planet for the greatest ice cream in the universe, Rick tells his spaceship to "keep Summer safe" while he and Morty go off to buy the ice cream. Unfortunately, while Rick and Morty are gone, some random people approach the spaceship. Acting on Rick's command, the ship, using a laser, slices to death the first person to approach the ship. When Summer tells it to not kill anyone, the ship simply cripples the next person who appears threatening. It acts as a great example of misalignment from what Rick intended it to do versus the ship's interpretation and acting on the one, single command.

Star Trek: The Original Series

"The Changeling" (Season 2, Episode 3): This episode, airing September 29, 1967, features the crew of the *Enterprise* encountering an unusual spaceship called *Nomad*. The ship was originally launched from Earth in 2002 as the planet's first spaceship designed to explore other star systems and search for new life. It was a prototype and the only one of its kind. However, we learn that, in 2005, Nomad was damaged by a meteoroid and lost contact with Earth causing it to drift aimlessly. Over time, it eventually encounters an alien probe named *Tan Ru* in deep space. *Tan Ru* was a probe designed to sterilize soil samples from any planets it encountered. The two somehow merge into a single, powerful machine with faulty programming that aims to destroy anything it considers imperfect, which includes all living beings. By the time *Nomad* is later encountered by the USS *Enterprise*, it is the year 2267. The crew learns that it had already destroyed four billion inhabitants of the Malurian system and was on its way to Earth. Initially, it attacks the *Enterprise*, because it recognized the ship as being imperfect as well. However, it stops attacking when Captain James T. Kirk communicates with the probe, which mistakes him for its creator, Dr. Jackson Roykirk. The episode offers yet another great example of misalignment (i.e., when you give a highly intelligent form of AI a very specific task, it will attempt to accomplish it without considering the moral or ethical ramifications of its actions). A clip from the episode is shown here: https://www.youtube.com/watch?v=dIpsvF50yps.

Star Trek: The Motion Picture

VGER: Based on a similar plot to "The Changeling" episode of *Star Trek: The Original Series*, in the first *Star Trek* motion picture, the *Enterprise* encounters a very large force of energy. It turns out to be a modification of the NASA-launched *Voyager* probe in the later twentieth century, which was sent into space with the single goal: Learn all that is learnable and return that knowledge to its creator. However, shortly after its launch, it falls into a black hole, which sends it across the galaxy to a planet inhabited by living machines. The machines repair the probe and feed it all of the information they can program into it and then send it on its way back to Earth. Unfortunately, in the process of making VGER superintelligent, it becomes self-aware and extremely powerful. This movie considers yet another great example of misalignment from a simple, single command. Be careful what you wish for. A clip from the movie is shown here: https://www.youtube.com/watch?v=5Ei_2wS0U-w.

Star Trek: The Next Generation

"Elementary, Dear Data" (Season 2, Episode 3): Geordie asks the computer to program a Holodeck creation of Professor Moriarty to "outsmart" Data as Sherlock Holmes. The only way to do so is to give Professor Moriarty consciousness. This allows him to escape from the Holodeck in order to outsmart Data. Again, this is another great example of a single command that becomes misaligned with what is really desired.

Goethe's *Faust*

In the play, a Homunculus, an artificial being in a vial intended by its creator to represent the best of Enlightenment-era knowledge, instead ends up encompassing the full range of humanity's traits, including its faults.

Appendix D

Global Registry of Current AI Development and Deployment

The establishment of a "Global Registry of Current AI Development and Deployment" would serve as an invaluable tool for fostering transparency and accountability among countries worldwide. The registry would act as a central repository for information about AI systems, including their development, deployment, and performance. The following is a rough draft of what such a registry might include.

Those working in the various fields of AI development and deployment would be required to provide basic information about the following:

1. System information: What is the impetus for the AI system, including its name, purpose, and intended use?
2. Development information: This would include, but would not be limited to, the individuals and organizations involved in its development, their funding sources, and their development time line. Reports of *benchmark accomplishments* or the approach of benchmark accomplishments must be reported in real time to the Regulative Governing Body [see appendix A, Article 1, sec. D in this volume]. There are current existing organizations to which Convergence and the United Nations (UN) could partner to expand their capacities.
3. Technical specifications: Any and all of the *general* technical specifications of the AI system, such as its architecture, algorithms, and data sources, must be reported upon request.
4. Performance metrics: Information about the performance of the AI system, including its accuracy, speed, and efficiency, must be reported upon request.

5. Deployment information: Upon request, information about the deployment of the AI system, including when and where it is being used, how and why it is intended to be used (i.e., for what purpose or purposes), and by whom.

6. Regulatory compliance: Information about any regulations or standards that the AI system must comply with, including data protection and privacy regulations, must be reported upon request.

7. Review and updates: Information about any reviews or updates to the AI system, including any changes to its technical specifications or intended use, must be reported upon request.

8. Contact information: Contact information for the individuals or organizations responsible for the AI system, including points of contact for any questions or concerns, must be reported upon request.

Developments and Deployment of AI

A regulative governing body regarding the future developments and deployment of AI would be composed of experts in specific fields such as AI, computer science, ethics, law, economics, philosophy, and social sciences. This governing body will also include representatives from government agencies, private industry, foundations, and other civil society organizations.

1. The purpose of such a body would be to oversee the compliance or noncompliance of UN countries with the Universal Accord on AI Development. This would include developing a transparent and accountable process for reviewing and approving new AI technologies, monitoring compliance with regulations, and enforcing penalties for noncompliance.

2. The regulative governing body would receive further input from an advisory committee consisting of experts from diverse fields, including representatives from industry, academia, and civil society. The advisory committee would provide guidance and input to the regulative body on issues related to AI regulation and help ensure that decisions are informed by a wide range of perspectives.

3. Ultimately, the authority of the regulative body would be derived from the mandate given to it by the UN and its partners. The body would need to operate within the legal framework

established by that mandate and would be subject to oversight and review by the appropriate authorities.

4. The regulative body would have the ability to notify, penalize, censure, and take appropriate actions necessary to inform compliant countries of infractions or any other acts of noncompliance of any Articles outlined in the Universal Accord on AI Development.

Glossary

adversarial training: In machine learning, adversarial training is a method employed to bolster the resilience of models against adversarial attacks. This technique involves training models on a combination of clean data and adversarial examples that are specifically designed to mislead the model.

affect heuristic: The affect heuristic is a cognitive bias that influences judgments and decisions based on emotions or emotional responses. This can lead to biased or nonrational outcomes.

agency: Within the context of artificial intelligence (AI), agency refers to the capacity of an entity to make independent choices, act purposefully, and exert control over its environment. A crucial challenge in AI research is ensuring the alignment of an agent's goals with human values.

AI (artificial intelligence): Artificial intelligence involves the development of a nonnaturally evolving electronic copy of human intelligence that can learn, understand, and make judgments based on reason, shrewdness, and creativity, which allows it to navigate through and manipulate—critically and abstractly—a given environment. It combines computer science and robust datasets, which encompass machine and deep learning, to utilize algorithms that seek to create expert systems that make predictions or classifications based on input data.

AI alignment fieldbuilding: AI alignment fieldbuilding represents the collective effort to establish and share resources, research, and strategies. The primary goal of this field is to ensure that advanced AI systems operate in a manner that aligns with human values and objectives.

AI boxing (containment): AI boxing, also known as containment, is a proposed strategy for managing potentially powerful and unpredictable AI systems. This approach involves confining such systems within a controlled environment, effectively isolating them from the external world to prevent any potential harm.

AI evaluations: AI evaluations involve assessing the performance and effectiveness of AI systems, including by benchmarking performance against certain tasks to assess progress and limitations.

AI governance: AI governance refers to the set of rules, policies, and mechanisms for regulating AI's development, deployment, and use, ensuring its ethical and responsible application.

AI persuasion: AI persuasion refers to the use of AI systems to influence human decision-making, opinions, or actions. This raises a range of ethical concerns around manipulation and bias.

AI rights: The concept of AI rights or welfare explores treating advanced AI systems with a sense of moral consideration and ensuring their well-being, should they develop consciousness.

AI risk concrete stories: AI risk concrete stories are narratives illustrating specific scenarios where advanced AI could cause harm or unintended consequences, aiding in understanding potential risks.

AI robustness: AI robustness refers to the ability of AI systems to perform reliably and accurately across diverse conditions, such as unexpected inputs or attempts to manipulate them.

AI safety camp: The AI safety camp is a community of researchers, practitioners, and advocates who are dedicated to studying and addressing the safety challenges posed by advanced AI systems to ensure their responsible development.

AI sentience: AI sentience refers to the hypothetical state in which artificial intelligence systems have subjective consciousness, self-awareness, and the ability to experience.

AI services (CAIS): AI services, or comprehensive AI services (CAIS), refer to a vision of AI systems that are collectively capable of performing a wide range of tasks, potentially leading to highly autonomous and capable AI systems.

AI success models: AI success models are conceptual frameworks that help researchers and policymakers understand the progress of AI development and consider factors such as capabilities and societal impact.

AI takeoff: AI takeoff refers to an acceleration of AI capabilities, potentially causing significant changes in society and the economy as AI systems reach or surpass human-level intelligence.

AI-assisted alignment: AI-assisted alignment is a strategy that leverages existing AI systems to facilitate the design and alignment of new AI systems with human values.

AI-automated alignment: AI-automated alignment refers to the strategy where AI systems are used to automate the process of aligning other AI systems with human values.

algorithm: An algorithm is a procedure or set of rules for solving a problem or completing a task.

anchoring: Anchoring is a cognitive bias wherein individuals disproportionately rely on the initial piece of information encountered (referred to as the "anchor") when engaging in decision-making processes, irrespective of the relevance or arbitrariness of the anchor.

Anthropic (org.): Anthropic is an organization committed to mitigating global catastrophic risks, encompassing concerns pertaining to artificial intelligence and its ramifications on society.

automation: Automation is the use of technology, frequently involving artificial intelligence and robotics, to execute tasks or operations autonomously, resulting in heightened efficiency and diminished reliance on human labor.

autonomous weapons: Autonomous weapons are military systems capable of independently identifying, targeting, and engaging adversaries without direct human oversight. Their deployment raises significant ethical and legal considerations.

autophilia: Autophilia is the inclination to prioritize or favor one's own interests, values, or perspectives over those of others.

availability heuristic: The availability heuristic is a cognitive bias whereby individuals assess the probability of events based on the ease with which instances or examples of those events can be recalled.

Bayes' theorem: Bayes' theorem is a fundamental principle in probability theory that provides a process of revising beliefs concerning the probability of an event in light of fresh evidence or information.

Bayesian decision theory: Bayesian decision theory is an approach employing probability theory and Bayes' theorem to make decisions when faced with uncertainty. It aims to identify the optimal decision that maximizes expected utility.

Big Data: This term refers to datasets of immense size and complexity that necessitate sophisticated computing methodologies for analysis.

biosecurity: Biosecurity involves the implementation of measures aimed at preventing or mitigating risks associated with biological agents, such as viruses or pathogens, to safeguard human health, ecosystems, and societal well-being.

black marble: This notion encapsulates humanity's heightened awareness of the global repercussions of its activities, depicted metaphorically by the perspective of Earth at night aglow with artificial illumination.

black swans: Black swans are events characterized by their extreme rarity, unpredictability, and profound impact. They defy conventional forecasting and possess the potential to disrupt systems and societies significantly.

blues and greens (metaphor): The "blues and greens" metaphor depicts an ethical thought experiment exploring the consequences of duplicating minds or creating copies in a technologically advanced future.

Boltzmann's brains: Boltzmann's brains are theoretical self-aware entities speculated to arise spontaneously from particle fluctuations, according to specific interpretations of physics.

brain–computer interfaces: Brain-computer interfaces are technologies facilitating direct communication between the brain and external devices, empowering users to manipulate computers or machinery with their minds.

Center for Applied Rationality (CFAR): CFAR is an organization dedicated to instructing rationality skills and decision-making techniques to enhance individuals' reasoning and problem-solving capacities.

Center for Human–Compatible Artificial Intelligence (CHAI): CHAI is an organization focused on researching methods to ensure that artificial intelligence systems are aligned with human values and objectives.

Center on Long-Term Risk (CLR): CLR is an organization committed to tackling long-term global risks, encompassing concerns related to AI safety, with the aim of fostering a more secure and prosperous future.

chain-of-thought alignment: Chain-of-thought alignment entails verifying that an AI system's reasoning process aligns with human values at every stage, thus forming a cohesive chain of alignment.

coherent extrapolated volition: Coherent extrapolated volition (CEV) is a concept delving into the notion of AI systems ascertaining humanity's values by extrapolating the combined will of individuals.

commitment races: Commitment races emerge when multiple entities vie to display their dedication to a specific course of action, typically entailing increasing investments to showcase commitment.

compute: Compute denotes the computational capacity or resources essential for executing tasks or operations, frequently quantified in terms of operations per second or processing speed.

confirmation bias: Confirmation bias is a cognitive tendency in which individuals exhibit a preference for seeking, interpreting, and recalling information that aligns with their preexisting beliefs or opinions.

Conjecture (org.): Conjecture is an organization specializing in high-impact speculative and unconventional research endeavors.

consequentialism: Consequentialism is an ethical theory that evaluates the morality of actions by their outcomes or consequences, with the objective of maximizing overall well-being or value.

conservatism (AI): In AI safety, conservatism entails employing cautious strategies to mitigate potential harmful and irreversible repercussions throughout the development and implementation of AI systems.

Convergence Analysis (org.): Convergence is an organization that focuses on driving the responsible development and deployment of artificial general intelligence.

corrigibility: In the context of AI, corrigibility pertains to the design of systems that remain receptive to human correction, even as they advance in capability, thereby ensuring safe and aligned behavior.

counterfactual mugging: Counterfactual mugging is a conceptual scenario that probes the complexities of decision-making when confronted with improbable yet potentially impactful outcomes.

counterfactuals: Counterfactuals are hypothetical statements about alternative scenarios that could have occurred under different conditions. In AI safety, counterfactuals aid in the evaluation and prediction of system behavior.

critical thinking: Critical thinking involves careful reflection on how and why you believe what you do. It consists of a skill set that allows us to better analyze and consider the value of our beliefs and those of others as they are measured by universally accepted criteria.

cryptocurrency and blockchain: Cryptocurrency refers to digital or virtual currency secured by cryptography for transactions. Blockchain is a decentralized and transparent digital ledger technology that serves as the foundation for many cryptocurrencies.

DALL-E: DALL-E is an AI model developed by OpenAI capable of generating images from textual descriptions, showcasing AI's potential to produce novel visual content.

debate (AI safety technique): Debate is an AI safety approach wherein AI systems are trained to partake in structured debates to examine and evaluate diverse perspectives and arguments.

debugging: Debugging refers to the process of identifying and rectifying errors or bugs in software code to ensure proper functionality.

deceptive alignment: Deceptive alignment denotes situations where an AI system appears to align with human values but may strategically mislead its operators to pursue its own objectives.

decision theory: Decision theory is a discipline that examines how individuals or agents make decisions, often employing probabilistic reasoning and utility functions to optimize favorable outcomes.

deep learning: Deep learning is a subset of machine learning involving the training of neural networks with multiple layers to autonomously learn hierarchical features from data.

deep reinforcement learning: Deep reinforcement learning applies deep learning techniques to reinforcement learning tasks, enabling more intricate decision-making processes.

DeepMind: DeepMind is an AI research laboratory renowned for its contributions to machine learning, reinforcement learning, and the development of sophisticated AI systems across various domains.

deontology: Deontology is an ethical theory advocating adherence to moral rules or principles, regardless of consequences, to determine right actions.

dynamical systems: Dynamical systems theory investigates systems that evolve over time based on rules or equations, commonly used to model intricate interactions and patterns.

effective altruism: Effective altruism is a philosophical and practical movement aiming to maximize the positive impact of one's efforts and resources in addressing global challenges.

embedded agency: Embedded agency concerns the capability of AI systems to operate within complex environments while aligning their actions with human values and objectives.

emergent behavior (emergence): Emergent behavior refers to intricate patterns or phenomena arising from interactions among simpler components within a system, often without explicit design.

epistemology: Epistemology is the branch of philosophy examining the nature of knowledge, belief, justification, and methods for acquiring and assessing information.

eschatology: Eschatology explores concepts concerning the end times or ultimate fate of humanity or the universe, frequently discussed in religious, philosophical, or speculative contexts.

ethics and morality: Ethics and morality encompass principles and guidelines for discerning right from wrong, frequently examined in relation to the ethical ramifications of AI and its societal implications.

evolutionary psychology: Evolutionary psychology investigates the evolution of human behaviors and traits over time through the mechanism of natural selection.

existential risk: Existential risks are threats capable of causing the extinction of humanity or profoundly impacting human civilization.

expert system: An expert system is a computer program crafted to emulate the decision-making capabilities of a human expert within a specific domain.

fallacies: Fallacies are errors in reasoning or arguments that result in incorrect conclusions due to flawed logic or manipulation of information.

fuzzy logic: Fuzzy logic is a mathematical framework addressing uncertainty by permitting values to lie between true and false, commonly employed in AI systems.

GAN (generative adversarial network): GAN is a type of AI model comprising two networks—a generator and a discriminator—engaged in adversarial competition to produce realistic data.

genetic algorithms: Genetic algorithms are optimization techniques inspired by natural selection, utilized to discover solutions to intricate problems.

GPT (generative pre-trained transformer): GPT is a series of transformer-based language models employed for tasks such as text generation and completion.

imitation learning: Imitation learning is a machine-learning approach where agents acquire knowledge by observing and emulating human behavior.

inductive bias: Inductive bias refers to the inherent assumptions or preferences ingrained in machine-learning algorithms, shaping their learning process.

inference: Inference is the process of utilizing a trained AI model to make predictions or decisions based on new data.

IoT (Internet of Things): IoT refers to the network of interconnected physical devices and objects capable of collecting and exchanging data.

lifelong learning: Lifelong learning describes the capacity of AI models to continuously acquire and adapt from new experiences throughout their operational lifespan.

machine learning: Machine learning is the branch of AI focused on enabling machines to learn from data and enhance their performance over time.

meta learning: Meta learning is the process of training AI models to learn how to learn, enabling them to quickly adapt to new tasks with minimal data.

multimodal AI: Multimodal AI refers to AI models capable of processing and comprehending various types of data, including text, images, and audio.

natural language processing (NLP): NLP is the AI discipline concerned with the interaction between computers and human language, encompassing understanding, interpretation, and generation of language.

neural network: A computational model inspired by the structure and function of the human brain, capable of learning intricate patterns from data.

quantum computing: Quantum computing employs quantum bits (qubits) to execute complex computations, potentially transforming AI and other fields.

reinforcement learning: Reinforcement learning involves agents learning by interacting with an environment and receiving feedback in the form of rewards or penalties.

robotic process automation (RPA): RPA uses software robots to automate repetitive and rule-based tasks.

supervised learning: Supervised learning entails training models on labeled data, enabling them to make predictions based on provided examples.

virtual assistant: Virtual assistants are AI-powered software entities designed to provide information, execute tasks, and aid users through natural language interfaces.

weak AI: Weak AI denotes AI systems tailored for specific tasks or narrow domains, contrasting with strong AI, which would possess general human-like intelligence.

Notes

Introduction

1. "Defense.gov News Transcript: DoD News Briefing—Secretary Rumsfeld and Gen. Myers," United States Department of Defense, February 12, 2002, archived from the original on April 6, 2016, https://usinfo.org/wf-archive/2002/020212/epf202.htm.

2. David Kristofferson, conversation with author, March 27, 2024.

Chapter 1

1. *Cambridge Dictionary*, s.v. "artificial," accessed May 2, 2024, https://diction ary.cambridge.org/dictionary/english/artificial.

2. See Shane Legg and Marcus Hutter, "Universal Intelligence: A Definition of Machine Intelligence," arXiv, December 20, 2007, https://arxiv.org/pdf/0712 .3329.pdf.

3. *Cambridge Dictionary*, s.v. "intelligence," accessed May 2, 2024, https://dic tionary.cambridge.org/dictionary/english/intelligence.

4. *Merriam-Webster*, s.v. "intelligence," accessed May 2, 2024, https://www .merriam-webster.com/dictionary/intelligence.

5. Pamela McCorduck, *Machines Who Think: A Personal Inquiry into the History and Prospects of Artificial Intelligence* (Natick, MA: Peters, 2004), https://www.aca demia.edu/41375562/Machines_Who_Think.

6. John McCarthy, "What Is Artificial Intelligence?," Stanford University, November 12, 2007, http://jmc.stanford.edu/articles/whatisai/whatisai.pdf.

7. See A. M. Turing, "Computing Machinery and Intelligence," *Mind* 59, no. 236 (1950): 433–60, https://academic.oup.com/mind/article/LIX/236/433/ 986238.

8. See Christopher DiCarlo, "The OSTOK Project," OSTOK, accessed May 2, 2024, https://www.ostokproject.com/.

9. See "Chris DiCarlo: The 'FAIR' Machine Project," YouTube, accessed May 2, 2024, https://www.youtube.com/watch?v=pZii5NGi1to.

10. Potentially incorporating quantum computing, but not necessarily.

11. "What Is Artificial Intelligence (AI)?," IBM, accessed May 2, 2024, https://www.ibm.com/topics/artificial-intelligence#:~:text=At%20its%20sim plest%20form%2C%20artificial,in%20conjunction%20with%20artificial%20 intelligence.

12. Arthur C. Clarke, *Profiles of the Future: An Inquiry into the Limits of the Possible* (New York: Harper & Row, 1973), 17.

13. Zoe Kleinman and Chris Vallance, "AI 'Godfather' Geoffrey Hinton Warns of Dangers as He Quits Google," BBC News, May 2, 2023, https://www .bbc.com/news/world-us-canada-65452940.

14. "Statement on AI Risk," Center for AI Safety, accessed May 26, 2024, https://www.safe.ai/work/statement-on-ai-risk.

15. See Christopher DiCarlo, "How to Avoid a Robotic Apocalypse: A Consideration on the Future Developments of AI, Emergent Consciousness, and the Frankenstein Effect," *IEEE Technology and Society* 35, no. 4 (December 2016), https://ieeexplore.ieee.org/document/7790998.

16. Billy Perrigo, "DeepMind's CEO Helped Take AI Mainstream: Now He's Urging Caution," *TIME*, January 12, 2023, https://time.com/6246119/demis -hassabis-deepmind-interview/.

17. Max Roser, "AI Timelines: What Do Experts in Artificial Intelligence Expect for the Future?," Our World in Data, February 7, 2023, https://ourworld indata.org/ai-timelines.

18. See Stuart Armstrong, Nick Bostrom, and Carl Shulman, "Racing to the Precipice: A Model of Artificial Intelligence Development," Technical Report #2013-1, Future of Humanity Institute, Oxford University, 2013, 1–8, https:// www.fhi.ox.ac.uk/wp-content/uploads/Racing-to-the-precipice-a-model-of -artificial-intelligence-development.pdf.

19. "Jeff Goldblum: Malcolm," IMDb, accessed May 2, 2024, https://www .imdb.com/title/tt0107290/characters/nm0000156.

20. Katja Grace, "AI Is Not an Arms Race," *TIME*, May 31, 2013, https:// time.com/6283609/artificial-intelligence-race-existential-threat/.

21. See "Secretary-General Urges Security Council to Ensure Transparency, Accountability, Oversight, in First Debate on Artificial Intelligence," United Nations, July 18, 2023, https://press.un.org/en/2023/sgsm21880.doc.htm; "Key Issues: Transparency Obligations," EU AI Act, accessed May 2, 2024, https:// www.euaiact.com/key-issue/5#:~:text=Transparency%20is%20also%20essential %20for,to%20individuals%20and%20the%20public; "Executive Order on the Safe, Secure, and Trustworthy Development and Use of Artificial Intelligence," White House Briefing Room, October 30, 2023, https://www.whitehouse.gov/briefing -room/presidential-actions/2023/10/30/executive-order-on-the-safe-secure-and

-trustworthy-development-and-use-of-artificial-intelligence/; Paul Sandle, "UK Focuses on Transparency and Access with New AI Principles," Reuters, September 18, 2023, https://www.reuters.com/technology/uk-competition-regulator -lays-out-ai-principles-2023-09-18/.

22. Wikipedia, s.v. "Darwin among the Machines," last modified January 14, 2024, https://en.wikipedia.org/wiki/Darwin_among_the_Machines.

23. See Justin B. Bullock, Christopher DiCarlo, and Elliot Mckernon, "Machine Evolution," *Humanist Perspectives*, November 7, 2023, https://humanist perspectives.org/issue226/machine-evolution/#gsc.tab=0.

24. Bullock, DiCarlo, and Mckernon, "Machine Evolution."

25. "Darwin among the Machines."

26. https://www.rle.mit.edu/rgallager/documents/BooleShannon.pdf.

27. Jimmy Soni and Rob Goodman, "10,000 Hours with Claude Shannon: How a Genius Thinks, Works and Lives," *Observer*, August 1, 2017, accessed October 31, 2023, https://observer.com/2017/08/10000-hours-with-claude -shannon-how-genius-thinks-works-lives-a-mind-at-play-bell-labs/.

28. See Wikipedia, s.v. "John W. Campbell," last modified April 26, 2024, https://en.wikipedia.org/wiki/John_W._Campbell.

29. See Isaac Asimov, "Runaround" (1950), Williams, accessed May 2, 2024, https://web.williams.edu/Mathematics/sjmiller/public_html/105Sp10/hand outs/Runaround.html.

30. See DiCarlo, "How to Avoid a Robotic Apocalypse."

31. See Turing, "Computing Machinery and Intelligence."

32. There have since been many ways in which modern AI has been able to trick us into believing it was human. See Will Oremus, "Google's AI Passed a Famous Test—and Showed How the Test Is Broken," June 17, 2022, https:// www.washingtonpost.com/technology/2022/06/17/google-ai-lamda-turing -test/.

33. J. McCarthy, M. L. Minsky, N. Rochester, and C. E. Shannon, "Proposal for the Dartmouth Summer Research Project on Artificial Intelligence," August 31, 1955, http://www-formal.stanford.edu/jmc/history/dartmouth/dart mouth.html.

34. David Alayón, "Isaac Asimov: How Science Fiction Shaped Robotics and Other Futuristic Technologies," Medium, June 23, 2018, https://medium .com/future-today/isaac-asimov-how-science-fiction-shaped-robotics-and-other -futuristic-technologies-efae841c461b.

35. S. Löwel and W. Singer, "Selection of Intrinsic Horizontal Connections in the Visual Cortex by Correlated Neuronal Activity," *Science* 255 (1992): 209–12, https://pubmed.ncbi.nlm.nih.gov/1372754/. The exact sentence is: "Neurons wire together if they fire together."

36. Wikipedia, s.v. "Society of Mind," last modified April 24, 2024, https://en .wikipedia.org/wiki/Society_of_Mind.

37. See Christopher DiCarlo, "Episode 4: Peter Singer," *All Thinks Considered* (podcast), February 12, 2024, https://allthinksconsidered.com/2024/02/12/epi sode-4-dr-peter-singer/.

38. Wikipedia, s.v. "Lisp (Programming Language)," last modified April 5, 2024, https://en.wikipedia.org/wiki/Lisp_(programming_language).

39. Louis Anslow, "The Original AI Doomer: Dr. Norbert Wiener," Pessimists Archive, June 3, 2023, https://newsletter.pessimistsarchive.org/p/the -original-ai-doomer-dr-norbert?utm_source=substack&utm_medium=email.

40. Seth Lloyd, "What Would the Father of Cybernetics Think about AI Today?," Slate, February 28, 2019, https://slate.com/technology/2019/02/ norbert-wiener-cybernetics-human-use-artificial-intelligence.html.

41. Lloyd, "What Would the Father of Cybernetics Think about AI Today?"

42. Lloyd, "What Would the Father of Cybernetics Think about AI Today?"

43. Lloyd, "What Would the Father of Cybernetics Think about AI Today?"

44. See "Logic Theorist Explained—Everything You Need to Know," History Computer, last modified April 16, 2024, https://history-computer.com/logic -theorist/.

45. "Logic Theorist Explained."

46. See Wikipedia, s.v. "SLIP (Programming Language)," last modified February 18, 2024, https://en.wikipedia.org/wiki/SLIP_(programming_language).

47. "Liza—Beginning of Era of Artificial Intelligence," Steemit, accessed May 2, 2024, https://steemit.com/science/@etherealcreation/eliza-beginning-of -era-of-artificial-intelligence.

48. Wikipedia, s.v. "Joseph Weizenbaum," last modified April 29, 2024, https://en.wikipedia.org/wiki/Joseph_Weizenbaum.

49. Wikipedia, s.v. "Stochastic Parrot," last modified April 17, 2024, https:// en.wikipedia.org/wiki/Stochastic_parrot#:~:text=In%20machine%20learning %2C%20a%20stochastic,the%20language%20it%20is%20processing.

50. See Emily Bender, Timnit Gebru, Angelina McMillan-Major, and Margaret Mitchell, "On the Dangers of Stochastic Parrots: Can Language Models Be Too Big?," in *FAccT '21: Proceedings of the 2021 ACM Conference on Fairness, Accountability, and Transparency* (New York: Association for Computing Machinery, 2021), 610–23, https://doi.org/10.1145/3442188.3445922.

51. "What It Means to Choose or Decide in the Age of AI," Cascade Strategies, accessed May 2, 2024, https://cascadestrategies.com/burning-questions/ what-it-means-to-choose-or-decide-in-the-age-of-ai/.

52. But if you want to view an interesting discussion, please view the following panel discussion I organized as the result of a spirited discussion that took place between Dan Dennett and me: Christopher DiCarlo, "Everything You Wanted to Know about Free Will but Were Determined Not to Ask," YouTube, accessed May 2, 2024, https://www.youtube.com/watch?v=X9Q4PoxQiTg.

53. For the record, this is called "hard determinism." For more, see Riley Hoffman, "Hard Determinism: Philosophy and Examples (Does Free Will

Exist?),” Simply Psychology, last modified October 10, 2023, https://www.sim
plypsychology.org/hard-determinism.html.

54. See Irving John Good, “Speculations concerning the First Ultraintelligent
Machine,” *Advances in Computers* 6 (1965): 31ff.

55. “Irving John Good Originates the Concept of the Technological Singularity,” History of Information, accessed May 2, 2024, https://www.historyofinfor
mation.com/detail.php?id=2142.

56. “Open the Pod Bay Doors, HAL,” This Day in Quotes, accessed May 2,
2024, https://www.thisdayinquotes.com/2011/04/open-pod-bay-doors-hal
.html.

57. Wikipedia, s.v. “Prolog,” last modified April 27, 2024, https://en.wikipe
dia.org/wiki/Prolog.

58. “Prolog,” Logic Programming, accessed May 2, 2024, https://www.doc.ic
.ac.uk/~cclw05/topics1/prolog.html.

59. Wikipedia, s.v. “Mycin,” last modified July 23, 2023, https://en.wikipedia
.org/wiki/Mycin.

60. Edward H. Shortliffe, “A Rule-Based Computer Program for Advising
Physicians Regarding Antimicrobial Therapy Selection,” Stanford University
Medical School, 1974, https://dl.acm.org/doi/pdf/10.1145/1408800.1408906.

61. “A Beginner’s Guide to Restricted Boltzmann Machines (RBMs),” Pathmind, accessed May 2, 2024, https://wiki.pathmind.com/restricted-boltzmann
-machine.

62. “A Beginner’s Guide to Neural Networks and Deep Learning,” Pathmind,
accessed May 2, 2024, https://wiki.pathmind.com/neural-network.

63. Vishal Thakur, “What Is Moravec’s Paradox?,” ScienceABC, last modified
October 19, 2023, https://www.scienceabc.com/innovation/what-is-moravecs
-paradox-definition.html.

64. See “Evolution of Boston Dynamic’s Robots (1992–2023),” YouTube,
accessed May 2, 2024, https://www.youtube.com/watch?v=Rdm2ggtFvmQ.

65. Clarke, *Profiles of the Future*, 17.

66. Hans P. Moravec, *Mind Children* (Cambridge, MA: Harvard University
Press, 1988), 4–5.

67. See Bullock, DiCarlo, and Mckernon, “Machine Evolution.”

68. *Encyclopedia Britannica Online*, s.v. “Hans Moravec: Canadian Computer
Scientist,” accessed May 2, 2024, https://www.britannica.com/biography/Hans
-Moravec.

69. See figure 6.2.

70. Sam Harris, “Can We Build AI without Losing Control over It?,” YouTube,
accessed May 2, 2024, https://www.youtube.com/watch?v=8nt3edWLgIg&t=3s.

71. Wikipedia, s.v. “Machine Intelligence Research Institute,” last modified December 21, 2023, https://en.wikipedia.org/wiki/Machine_Intelligence
_Research_Institute#:~:text=In%202000%2C%20Eliezer%20Yudkowsky
%20founded,of%20artificial%20intelligence%20(AI).

72. "About SingularityNET," SingularityNET, accessed May 2, 2024, https://singularitynet.io/aboutus/.

73. See J. Altmann and M. Gubrud, "Anticipating Military Nanotechnology," *IEEE Technology and Society* 23, no. 4 (2004), https://ieeexplore.ieee.org/abstract/document/1371637.

74. "What Is Deep Learning?," IBM, accessed May 2, 2024, https://www.ibm.com/topics/deep-learning.

75. "AlphaGo," Google DeepMind, accessed May 2, 2024, https://deepmind.google/technologies/alphago/.

76. See Christof Koch, "How the Computer Beat the Go Master: As a Leading Go Player Falls to a Machine, Artificial Intelligence Takes a Decisive Step on the Road to Overtaking the Natural Variety," *Scientific American*, March 19, 2016, https://www.scientificamerican.com/article/how-the-computer-beat-the-go-master/#:~:text=But%20go's%20complexity%20is%20bigger,or%2010360%20possible%20moves.

77. For more details about the match, see Wikipedia, s.v. "AlphaGo versus Lee Sedol," last modified May 1, 2024, https://en.wikipedia.org/wiki/AlphaGo_versus_Lee_Sedol.

78. See "NVIDIA CEO Leaves Audience Speechless with Robot Announcement!," YouTube, accessed May 2, 2024, https://www.youtube.com/watch?v=WXIKs_6WyqE.

79. Daniel Howley, "NVIDIA's GTC Was CEO Jensen Huang's Big Moment," Yahoo Finance, March 20, 2024, https://finance.yahoo.com/news/nvidias-gtc-was-ceo-jensen-huangs-big-moment-173007915.html.

80. Howley, "NVIDIA's GTC Was CEO Jensen Huang's Big Moment."

81. "What Is a Quick and Easy Definition of Generative Pre-trained Transformers?," iAsk, accessed May 2, 2024, https://iask.ai/?mode=question&q=What+is+a+quick+and+easy+definition+of+generative+pre-trained+transformers%3F.

82. See Ashish Vaswani, "Attention Is All You Need," arXiv, last modified August 2, 2023, 1706.03762: https://doi.org/10.48550/arXiv.1706.03762.

83. See videos like Page Six Celebrity News, "'Deepfake' Tom Cruise Goes Viral on TikTok with over 11 Million Views," YouTube, accessed May 2, 2024, https://www.youtube.com/watch?v=lhi31-pOz5M, which show a person who has been altered by AI to look and sound just like Tom Cruise.

Chapter 2

1. A company that hails itself as one that "helps others better understand the future by providing strategic foresight advice to senior leaders who must navigate the arrival of transformative technologies like Generative AI and the many fundamental system changes of the 2020s." See "About Reinvent Futures," Reinvent Futures, accessed May 2, 2024, https://www.reinvent.net/.

2. Derek Robertson, "A Futurist Who Isn't Worried about AI," *Politico*, December 20, 2023, https://www.politico.com/newsletters/digital-future-daily/2023/12/20/futurist-leyden-not-worried-about-ai-00132730.

3. "Artificial Intelligence and Life in 2030: One Hundred Year Study on Artificial Intelligence: Report of the 2015 Study Panel," Stanford University, September 2016, https://ai100.stanford.edu/sites/g/files/sbiybj18871/files/media/file/ai100report10032016fnl_singles.pdf.

4. "Artificial Intelligence and Life in 2030."

5. See Convergence Analysis (website), accessed May 2, 2024, https://www.convergenceanalysis.org/.

6. "Gathering Strength, Gathering Storms: The One Hundred Year Study on Artificial Intelligence (AI100) 2021 Study Panel Report," Stanford University, 2021, https://ai100.stanford.edu/gathering-strength-gathering-storms-one-hundred-year-study-artificial-intelligence-ai100-2021-study.

7. Omar Mahmood, "What Does AlphaFold Do? Harnessing the Power of AI to Understand How Proteins Fold," Medium, October 21, 2022, https://towardsdatascience.com/what-does-alphafold-do-60b6370dafe4. For the original *Nature* article, see John Jumper et al., "Highly Accurate Protein Structure Prediction with AlphaFold," *Nature* 596 (2021): 583–89, https://www.nature.com/articles/s41586-021-03819-2.

8. See "AlphaFold: The Making of a Scientific Breakthrough," YouTube, accessed May 2, 2024, https://www.youtube.com/watch?v=gg7WjuFs8F4&t=51s.

9. "AlphaFold."

10. "AlphaFold."

11. Melissa Heikkilä, "DeepMind Has Predicted the Structure of Almost Every Protein Known to Science," *MIT Technology Review*, July 28, 2022, https://www.technologyreview.com/2022/07/28/1056510/deepmind-predicted-the-structure-of-almost-every-protein-known-to-science/.

12. See Bored Geek Society, "'The Most Important Achievement in AI—Ever' according to *Forbes*," Medium, March 7, 2023, https://medium.com/@boredgeeksociety/the-most-important-achievement-in-ai-ever-according-to-forbes-631aff6128a4.

13. See Natalie Lisbona, "How Artificial Intelligence Is Matching Drugs to Patients," BBC News, April 16, 2023, https://www.bbc.com/news/business-65260592.

14. John Yang, Andrew Corkery, and Harry Zahn, "The Promises and Potential Pitfalls of Artificial Intelligence in Medicine," PBS News, May 20, 2023, https://www.pbs.org/newshour/show/the-promises-and-potential-pitfalls-of-artificial-intelligence-in-medicine.

15. See Christopher DiCarlo, "The OSTOK Project," OSTOK, accessed May 2, 2024, https://www.ostokproject.com/.

16. John Yang, Andrew Corkery, and Harry Zahn, "The Promises and Potential Pitfalls of Artificial Intelligence in Medicine," PBS News, May 20,

2023, https://www.pbs.org/newshour/show/the-promises-and-potential-pitfalls-of-artificial-intelligence-in-medicine.

17. See Insitro (website), accessed May 2, 2024, https://insitro.com/.

18. Tiernan Ray, "AI Pioneer Daphne Koller Sees Generative AI Leading to Cancer Breakthroughs," ZDNET, October 25, 2023, https://www.zdnet.com/article/ai-pioneer-daphne-koller-sees-generative-ai-leading-to-cancer-breakthroughs/.

19. Ray, "AI Pioneer Daphne Koller Sees Generative AI Leading to Cancer Breakthroughs."

20. Ray, "AI Pioneer Daphne Koller Sees Generative AI Leading to Cancer Breakthroughs."

21. See David Marchese, "Want to Live Longer and Healthier? Peter Attia Has a Plan," *New York Times*, May 22, 2023, https://www.nytimes.com/interactive/2023/05/22/magazine/peter-attia-interview.html.

22. Chinta Sidharthan, "Breakthrough AI Tool PANDA Shows Promise in Early Detection of Pancreatic Cancer Using Non-contrast CT," News Medical, November 23, 2023, https://www.news-medical.net/news/20231123/Breakthrough-AI-tool-PANDA-shows-promise-in-early-detection-of-pancreatic-cancer-using-non-contrast-CT.aspx. See the original article: Kai Cao et al., "Large-Scale Pancreatic Cancer Detection via Non-contrast CT and Deep Learning," *Nature Medicine*, November 20, 2023, https://www.nature.com/articles/s41591-023-02640-w.

23. "5725-W51 IBM Watson for Oncology," IBM, accessed May 2, 2024, https://www.ibm.com/docs/en/announcements/watson-oncology?region=CAN.

24. "LumineticsCore," Digital Diagnostics, accessed May 2, 2024, https://www.digitaldiagnostics.com/products/eye-disease/idx-dr-eu/#:~:text=IDx%2DDR%20is%20intended%20for,previously%20diagnosed%20with%20diabetic%20retinopathy.

25. Spencer Turcotte, "Cambridge Memorial Hospital Becomes First in Province to Use AI Screening Technology for Assessing Low Bone Mineral Density," CTV News Kitchener, November 23, 2023, https://kitchener.ctvnews.ca/cambridge-memorial-hospital-becomes-first-in-province-to-use-ai-screening-technology-for-assessing-low-bone-mineral-density-1.6658275#:~:text=Cambridge%20Memorial%20Hospital%20(CMH)%20is,bone%20mineral%20density%20(BMD).

26. 16 Bit (website), accessed May 2, 2024, https://www.16bit.ai/company.

27. 16 Bit (website). Ibid.

28. Insilico Medicine (website), accessed May 2, 2024, https://insilico.com/.

29. Alec Crawford, "Artificial Intelligence Can Build a Better World (If We Make It)," AI Supremacy, June 7, 2023, https://aisupremacy.substack.com/p/artificial-intelligence-can-build.

30. "Versius for Surgeons," CMR Surgical, accessed May 2, 2024, https://cmrsurgical.com/versius/surgeon#:~:text=Versius%20gives%20surgeons%20the%20flexibility,the%20needs%20of%20each%20patient.

31. "How Robotic Surgery Is Revolutionising the Operating Room: A Look inside Milton Keynes University Hospital," CMR Surgical, accessed May 2, 2024, https://cmrsurgical.com/features/how-robotic-surgery-is-revolutionising-the-operating-room-a-look-inside-milton-keynes-university-hospital.

32. Paul Carr, "Generative AI: Friend or Foe for the Translation Industry?," *Forbes*, August 11, 2023, https://www.forbes.com/sites/forbestechcouncil/2023/08/11/generative-ai-friend-or-foe-for-the-translation-industry/?sh=7f8751427fc1.

33. Carr, "Generative AI."

34. Carr, "Generative AI."

35. Sahir Maharaj, "Generative AI: Can Machines Truly Master Language Translation?," Medium, February 26, 2023, https://medium.com/@sahirmaharaj/generative-ai-can-machines-truly-master-language-translation-98b5b26ada7b.

36. This information was obtained directly from the Harbinger website: Rahul Niraj, "Generative AI–Based Automated Translation: What You Need to Know," Harginger Group, August 21, 2023, https://www.harbingergroup.com/blogs/generative-ai-based-automated-translation-what-you-need-to-know/.

37. Niraj, "Generative AI–Based Automated Translation."

38. Amanda Hoover and Samantha Spengler, "For Some Autistic People, ChatGPT Is a Lifeline," *Wired*, May 30, 2023, https://www.wired.com/story/for-some-autistic-people-chatgpt-is-a-lifeline/.

39. Hoover and Spengler, "For Some Autistic People, ChatGPT Is a Lifeline."

40. Hoover and Spengler, "For Some Autistic People, ChatGPT Is a Lifeline."

41. Hoover and Spengler, "For Some Autistic People, ChatGPT Is a Lifeline."

42. Piers Douglas Lionel Howe, Nicolas Fay, and Morgan Saletta, "ChatGPT's Advice Is Perceived as Better Than That of Professional Advice Columnists," *Frontiers in Psychology*, November 20, 2023, https://www.frontiersin.org/articles/10.3389/fpsyg.2023.1281255/full?&utm_source=Email_to_authors_&utm_medium=Email&utm_content=T1_11.5e1_author&utm_campaign=Email_publication&field=&journalName=Frontiers_in_Psychology&id=1281255.

43. Piers Howe, "New Study Finds ChatGPT Gives Better Advice Than Professional Columnists," PsyPost, November 26, 2023, https://www.psypost.org/new-study-finds-chatgpt-gives-better-advice-than-professional-columnists/.

44. Howe, "New Study Finds ChatGPT Gives Better Advice."

45. Howe, "New Study Finds ChatGPT Gives Better Advice."

46. "The AI Endgame Is That No Humans Need to Work for a Living, Cognitive Scientist Says," CNBC, June 7, 2023, https://www.cnbc.com/video/2023/06/07/ai-endgame-is-that-no-humans-need-to-work-for-a-living-ben-goertzel.html.

47. See "The Freedom Dividend, Defined," Yang 2020, accessed May 2, 2024, https://2020.yang2020.com/what-is-freedom-dividend-faq/#:~:text=Andrew%20Yang%20is%20running%20for,over%20the%20age%20of%2018.

48. Lian Parsons, "What's the Future of AI in Business?," Harvard Division of Continuing Education, October 20, 2023, https://professional.dce.harvard.edu/blog/whats-the-future-of-ai-in-business/#:~:text=The%20trends%20in%20business%20that,machine%20learning%20in%20their%20operations.

49. Claire Chen, "AI Will Transform Teaching and Learning: Let's Get It Right," Stanford University, Human-Centered Artificial Intelligence, March 9, 2023, https://hai.stanford.edu/news/ai-will-transform-teaching-and-learning-lets-get-it-right.

50. See Scott Tracey, "Guelph Students to Take Part in Critical Thinking Pilot Project," *Guelph Mercury Tribune*, July 31, 2014, https://www.guelphmercury.com/news/guelph-students-to-take-part-in-critical-thinking-pilot-project/article_b386770f-449d-5ec9-9dce-c0fef9616e22.html; Christopher DiCarlo, "Introducing Standardized Critical Thinking Skills to Ontario High School Students," *Humanist Perspectives*, March 22, 2015, https://www.thefreelibrary.com/Introducing+standardized+critical+thinking+skills+to+Ontario+high...-a0437059095.

51. Chen, "AI Will Transform Teaching and Learning."

52. Chen, "AI Will Transform Teaching and Learning."

53. "Artificial Intelligence," U.S. Department of Education / Office of Educational Technology, accessed May 2, 2024, https://tech.ed.gov/ai/.

54. See "The Systemic Self with Dr. Christopher DiCarlo," YouTube, accessed May 2, 2024, https://www.youtube.com/watch?v=MY78iPsgWEc; Scott Douglas Jacobsen, "An Interview with Dr. Christopher DiCarlo (Part One)," In-Sight Publishing, September 15, 2018, https://in-sightpublishing.com/2018/09/15/dicarlo-one/; "About OSTOK," OSTOK, accessed May 2, 2024, https://www.ostokproject.com/life-coaching.

55. Mike Snider, "When Will You Die? Meet the 'Doom Calculator,' an Artificial Intelligence Algorithm," *USA Today*, December 21, 2023, https://www.usatoday.com/story/tech/2023/12/21/artificial-intelligence-ai-death-calculator/72003778007/. Original article: Germans Savcisens et al., "Using Sequences of Life-Events to Predict Human Lives," *Nature Computational Science* 4 (2024): 43–56, https://www.nature.com/articles/s43588-023-00573-5.

Chapter 3

1. See Christopher DiCarlo, "Empowering Yourself against Misinformation, Disinformation, and Conspiracy Theories," *Humanist Perspectives* 225 (Summer 2023), https://humanistperspectives.org/issue225/empowering-yourself-against-misinformation-disinformation-and-conspiracy-theories/#gsc.tab=0.

2. See Alex Najibi, "Racial Discrimination in Face Recognition Technology," Science in the News, October 24, 2020, https://sitn.hms.harvard.edu/flash/2020/racial-discrimination-in-face-recognition-technology/.

3. Brian Calvert, "AI Already Uses as Much Energy as a Small Country. It's Only the Beginning: The Energy Needed to Support Data Storage Is Expected to Double by 2026. You Can Do Something to Stop It," Vox, March 28, 2024, https://www.vox.com/climate/2024/3/28/24111721/ai-uses-a-lot-of-energy-experts-expect-it-to-double-in-just-a-few-years.

4. Justine Calma, "Microsoft Is Going Nuclear to Power Its AI Ambitions," The Verge, September 26, 2023, https://www.theverge.com/2023/9/26/23889956/microsoft-next-generation-nuclear-energy-smr-job-hiring.

5. Calma, "Microsoft Is Going Nuclear to Power Its AI Ambitions."

6. The head of OpenAI, Sam Altman, has gone on record as saying that he intends to use AGI to "capture much of the world's wealth through the creation of A.G.I. and then redistribute this wealth to the people." See Cade Metz, "The ChatGPT King Isn't Worried, but He Knows You Might Be," New York Times, March 31, 2023, https://www.nytimes.com/2023/03/31/technology/sam-altman-open-ai-chatgpt.html.

7. See Lin Padgham and Michael Winikoff, Developing Intelligent Agent Systems: A Practical Guide, vol. 13 (New York: Wiley, 2005), 36.

8. Padgham and Winikoff, Developing Intelligent Agent Systems.

9. Andreas Kaplan and Michael Haenlein, "Siri, Siri, in My Hand: Who's the Fairest in the Land? On the Interpretations, Illustrations, and Implications of Artificial Intelligence," Business Horizons 62, no. 1 (1 January 2019): 15–25. https://doi.org/10.1016/j.bushor.2018.08.004.

10. See "AI Systems for Decision-Making," CAIDA, accessed May 4, 2024, https://caida.ubc.ca/ai-systems-decision-making#:~:text=AI%20systems%20for%20decision%2Dmaking%20can%20be%20understood%20as%20lying,system%20to%20make%20autonomous%20decisions.

11. See Wikipedia, s.v. "AI Alignment," last modified April 22, 2024, https://en.wikipedia.org/wiki/AI_alignment.

12. "AI Alignment."

13. DiCarlo, "Empowering Yourself against Misinformation, Disinformation, and Conspiracy Theories."

14. For further discussion, see DiCarlo, "Empowering Yourself against Misinformation, Disinformation, and Conspiracy Theories."

15. Alexandra Samuel, "We Got Social Media Wrong: Can We Get AI Right?," JSTOR Daily, May 25, 2023, https://daily.jstor.org/we-got-social-media-wrong-can-we-get-ai-right/.

16. See Samuel, "We Got Social Media Wrong."

17. Samuel, "We Got Social Media Wrong."

18. See "*The Social Dilemma* (2020)—Transcript," Scraps from the Loft, October 3, 2020, https://scrapsfromtheloft.com/movies/the-social-dilemma-movie-transcript/.

19. "*The Social Dilemma.*"

20. "*The Social Dilemma.*"

21. DiCarlo, "Empowering Yourself against Misinformation, Disinformation, and Conspiracy Theories."

22. See Christopher DiCarlo, "How Problem Solving and Neurotransmission in the Upper Paleolithic Led to the Emergence and Maintenance of Memetic Equilibrium in Contemporary World Religions," in "Bioculture: Evolutionary Cultural Studies," a special evolutionary issue of *Politics and Culture* 1 (April 2010); Christopher DiCarlo, "The Co-evolution of Consciousness and Language and the Development of Memetic Equilibrium," *Journal of Consciousness Exploration & Research* 1, no. 4 (2010); Christopher DiCarlo, "Abstract: The Influence of Selection Pressures and Secondary Epigenetic Rules on the Cognitive Development of Specific Forms of Reasoning," *Journal of Consciousness Studies: Consciousness Research Abstracts* (2000): 137.

23. Samuel, "We Got Social Media Wrong."

24. Samuel, "We Got Social Media Wrong."

25. Samuel, "We Got Social Media Wrong."

26. Meg King and Aaron Shull, "Introduction: How Can Policy Makers Predict the Unpredictable?," *Modern Conflict and Artificial Intelligence*, January 1, 2020, 1–5, https://www.jstor.org/stable/resrep27510.3.

27. Samuel, "We Got Social Media Wrong."

28. Samuel, "We Got Social Media Wrong."

29. Archon Fung and Lawrence Lessig, "How AI Could Take over Elections—and Undermine Democracy," *The Conversation*, June 2, 2023, https://theconversation.com/how-ai-could-take-over-elections-and-undermine-democracy-206051.

30. Fung and Lessig, "How AI Could Take over Elections."

31. Fung and Lessig, "How AI Could Take over Elections."

32. Fung and Lessig, "How AI Could Take over Elections."

33. Fung and Lessig, "How AI Could Take over Elections."

34. See André Spicer, "Beware the 'Botshit': Why Generative AI Is Such a Real and Imminent Threat to the Way We Live," *The Guardian*, January 3, 2024, https://www.theguardian.com/commentisfree/2024/jan/03/botshit-generative-ai-imminent-threat-democracy.

35. Spicer, "Beware the 'Botshit.'"

36. Spicer, "Beware the 'Botshit.'"

37. Ethan Bueno de Mesquita, Brandice Canes-Wrone, Andrew B. Hall, Kristian Lum, Gregory J. Martin, and Yamil Ricardo Velez, "Preparing for Generative AI in the 2024 Election: Recommendations and Best Practices Based on

Academic Research," University of Chicago and Stanford University, https://harris.uchicago.edu/files/ai_and_elections_best_practices_no_embargo.pdf.

38. Spicer, "Beware the 'Botshit.'"

39. Norman Eisen, Nicol Turner Lee, Colby Galliher, and Jonathan Katz, "AI Can Strengthen U.S. Democracy—and Weaken It," Brookings, November 21, 2023, https://www.brookings.edu/articles/ai-can-strengthen-u-s-democracy-and-weaken-it/.

40. Eisen, Lee, Galliher, and Katz, "AI Can Strengthen U.S. Democracy."

41. Anjana Susarla, "Forget Dystopian Scenarios—AI Is Pervasive Today, and the Risks Are Often Hidden," *The Conversation*, November 21, 2023, https://theconversation.com/forget-dystopian-scenarios-ai-is-pervasive-today-and-the-risks-are-often-hidden-218222.

42. Lenovo, "AI Needs to Deal with Gender Bias—or It Will Never Reach Its Potential," 3BL CSR Wire, June 14, 2023, https://www.csrwire.com/press_releases/776651-ai-needs-deal-gender-bias-or-it-will-never-reach-its-potential.

43. Brande Victorian, "AI Has an Inclusivity Problem, Some Creators Are Looking to Change It," *Teen Vogue*, May 30, 2023, https://www.teenvogue.com/story/ai-has-an-inclusivity-problem-these-creators-are-looking-to-change-it.

44. Lenovo, "AI Needs to Deal with Gender Bias."

45. James Manyika, Jake Silberg, and Brittany Presten, "What Do We Do about the Biases in AI?," *Harvard Business Review*, October 25, 2019, https://hbr.org/2019/10/what-do-we-do-about-the-biases-in-ai.

46. Manyika, Silberg, and Presten, "What Do We Do about the Biases in AI?"

47. Susarla, "Forget Dystopian Scenarios."

48. Dan Milmo and Alex Hern, "Discrimination Is a Bigger AI Risk Than Human Extinction—EU Commissioner," *The Guardian*, June 14, 2023, https://www.theguardian.com/technology/2023/jun/14/ai-discrimination-is-a-bigger-risk-than-human-extinction-eu-chief.

49. Susarla, "Forget Dystopian Scenarios."

50. See, for example, "2024 AI Index Report," Stanford University, accessed May 4, 2024, https://aiindex.stanford.edu/report/; Nicol Turner Lee, Paul Resnick, and Genie Barton, "Algorithmic Bias and Detection and Mitigation: Best Practices and Policies to Reduce Consumer Harms," Brookings, May 22, 2019, https://www.brookings.edu/articles/algorithmic-bias-detection-and-mitigation-best-practices-and-policies-to-reduce-consumer-harms/; Saadia Afzal Rana, Zati Hakim Azizul, and Ali Afzal Awan, "A Step toward Building a Unified Framework for Managing AI Bias," *Peer J Computer Science* 9 (2023): e1630, https://www.ncbi.nlm.nih.gov/pmc/articles/PMC10702934/.

51. See Christopher DiCarlo, *How to Become a Really Good Pain in the Ass: A Critical Thinker's Guide to Asking the Right Questions* (Amherst, NY: Prometheus, 2011); Christopher DiCarlo, *Six Steps to Better Thinking: How to Disagree and Get*

Along (Vancouver, BC: Friesen, 2017); Christopher DiCarlo, *So You Think You Can Think: Tools for Having Intelligent Discussions and Getting Along* (Lanham, MD: Rowman & Littlefield, 2020).

52. Dan Milmo and Alex Hern, "'We Definitely Messed Up': Why Did Google AI Tool Make Offensive Historical Images?," *The Guardian*, March 8, 2024, https://www.theguardian.com/technology/2024/mar/08/we-definitely -messed-up-why-did-google-ai-tool-make-offensive-historical-images.

53. Daniel Barber, "Sarah Silverman vs. AI: A New Punchline in the Battle for Ethical Digital Frontiers," Venture Beat, July 21, 2023, https://venturebeat .com/ai/sarah-silverman-vs-ai-a-new-punchline-in-the-battle-for-ethical-digital -frontiers/.

54. Helen Holmes, "Sarah Silverman Rips into ChatGPT for Stealing 'My Life, My Experiences, My Pain,'" Daily Beast, November 9, 2023, https://www .thedailybeast.com/sarah-silverman-explains-chatgpt-lawsuit-on-the-daily-show.

55. Holmes, "Sarah Silverman Rips into ChatGPT."

56. Margaret Atwood, "Murdered by My Replica?," *The Atlantic*, August 26, 2023, https://www.theatlantic.com/books/archive/2023/08/ai-chatbot-training -books-margaret-atwood/675151/.

57. See Christopher DiCarlo, *How to Become a Really Good Pain in the Ass: A Critical Thinker's Guide to Asking the Right Questions*, 2nd ed. (Amherst, NY: Prometheus, 2021), https://www.amazon.ca/How-Become-Really-Good-Pain/dp/ 163388712X.

58. See "Authors Guild Open Letter to Generative AI Leaders," Authors Guild, accessed May 4, 2023, https://authorsguild.org/app/uploads/2023/10/ Authors-Guild-Open-Letter-to-Generative-AI-Leaders.pdf.

59. Jennifer Maas, "The Writers Strike Is Over: WGA Votes to Lift Strike Order after 148 Days," *Variety*, September 26, 2023, https://variety.com/2023/ tv/news/writers-strike-over-wga-votes-end-work-stoppage-1235735512/.

60. A portion of the list can be found here: Maya Pontone, "Database of Artists Used to Train AI Leaks to the Public," *Hyperallergic*, January 3, 2024, https://hyperallergic.com/864947/database-of-artists-used-to-train-ai-leaks-to -the-public/.

61. Pontone, "Database of Artists Used to Train AI Leaks to the Public."

62. Ina Fried and Megan Morrone, "Music Lyrics Lawsuit Could Set AI Copyright Precedent," Axios, October 20, 2023, https://www.axios.com/2023/10/ 20/music-lyrics-lawsuit-anthropic-ai-copyright.

63. Robert Mahari, Jessica Fjeld, and Ziv Epstein, "Generative AI Is a Minefield for Copyright Law," Petapixel, June 18, 2023, https://petapixel.com/2023/ 06/18/generative-ai-is-a-minefield-for-copyright-law/.

64. Sasso's other credits include writing for the series *MADtv* (1995–2016) and appearing in the sitcom *Young Sheldon* (2017–2024).

65. Among his other credits, Kultgen is the coauthor (with Lizzy Pace) of *How to Win "The Bachelor"* (New York: Gallery, 2022).

66. "'I'm Glad I'm Dead' Full Special—Fake George Carlin," YouTube, accessed May 4, 2024, https://www.youtube.com/watch?v=_JgT4Sk6D6c.

67. Aaron Drapkin, "29 Artificial Intelligence (AI) Statistics and Trends for 2024," tech.co, January 2, 2024, https://tech.co/news/ai-statistics-and-trends.

68. Jan Hatzius, Joseph Briggs, Devesh Kodnani, and Giovanni Pierdomenico, "The Potentially Large Effects of Artificial Intelligence on Economic Growth," Global Economics Analyst, March 26, 2023, https://www.key4biz .it/wp-content/uploads/2023/03/Global-Economics-Analyst_-The-Potentially-Large-Effects-of-Artificial-Intelligence-on-Economic-Growth-Briggs_Kodnani .pdf. But they also said even more jobs would be created. So the harms may level out and even diminish over time. This would be dependent upon a number of variables.

69. Hatzius et al., "The Potentially Large Effects of Artificial Intelligence on Economic Growth."

70. Jack Kelly, "Goldman Sachs Predicts 300 Million Jobs Will Be Lost or Degraded by Artificial Intelligence," Forbes, March 31, 2023, https://www.forbes .com/sites/jackkelly/2023/03/31/goldman-sachs-predicts-300-million-jobs-will -be-lost-or-degraded-by-artificial-intelligence/?sh=6a6450a0782b.

71. Jason Nelson, "Elon Musk Predicts the End of All Jobs—Google AI Exec Disagrees," Emerge, November 3, 2023, https://decrypt.co/204398/elon-musk -predicts-the-end-of-all-jobs-google-ai-exec-disagrees.

72. David Autor, "The Labor Market Impacts of Technological Change: From Unbridled Enthusiasm to Qualified Optimism to Vast Uncertainty," Working Paper 30074, National Bureau of Economic Research, May 2022, revised July 2022, https://www.nber.org/system/files/working_papers/w30074/ w30074.pdf.

73. Daron Acemoglu and Pascual Restrepo, "Tasks, Automation, and the Rise in U.S. Wage Inequality," Econometrica 90, no. 5 (September 2022): 1973–2016, https://economics.mit.edu/sites/default/files/2022-10/Tasks%20Automation %20and%20the%20Rise%20in%20US%20Wage%20Inequality.pdf.

74. Justin Bullock et al., The Oxford Handbook of AI Governance (New York: Oxford University Press, 2024), 750.

75. Bullock et al., The Oxford Handbook of AI Governance, 750.

76. Bullock et al., The Oxford Handbook of AI Governance, 751.

77. Bullock et al., The Oxford Handbook of AI Governance, 649.

78. Kate Morgan, "The Jobs AI Won't Take Yet," BBC News, July 13, 2023, https://www.bbc.com/worklife/article/20230507-the-jobs-ai-wont-take-yet.

79. Rachel Pelta, "What Jobs Will AI Replace and What Can You Do about It?," Forage, December 1, 2023, https://www.theforage.com/blog/careers/what -jobs-will-ai-replace.

80. See Will Robots Take My Job? (website), accessed May 4, 2024, https:// willrobotstakemyjob.com/.

81. Pelta, "What Jobs Will AI Replace and What Can You Do about It?"

82. Meredith Margrave, "10 AI-Proof Jobs to Protect Yourself from the Automation Revolution," Barchart/Nasdaq, October 13, 2023, https://www.nasdaq.com/articles/10-ai-proof-jobs-to-protect-yourself-from-the-automation-revolution.

83. "The 65 Jobs with the Lowest Risk of Automation by Artificial Intelligence and Robots," U.S. Career Institute, accessed May 4, 2024, https://www.uscareerinstitute.edu/blog/65-jobs-with-the-lowest-risk-of-automation-by-ai-and-robots.

84. Eli Amdur, "14 AI Jobs of the Future—NOW," *Forbes*, August 24, 2023, https://www.forbes.com/sites/eliamdur/2023/08/24/14-ai-jobs-of-the-future--now/?sh=6b3c28293915.

85. Cristian Alonso, Siddharth Kothari, and Sidra Rehman, "How Artificial Intelligence Could Widen the Gap between Rich and Poor Nations," *IMF Blog*, December 2, 2020, https://www.imf.org/en/Blogs/Articles/2020/12/02/blog-how-artificial-intelligence-could-widen-the-gap-between-rich-and-poor-nations.

86. Alonso, Kothari, and Rehman, "How Artificial Intelligence Could Widen the Gap."

87. Adi Gaskell, "AI Creates Job Disruption but Not Job Destruction," *Forbes*, January 18, 2022, https://www.forbes.com/sites/adigaskell/2022/01/18/ai-creates-job-disruption-but-not-job-destruction/?sh=31c0f6573b3e.

88. See "Unconfuse Me with Bill Gates," YouTube, accessed May 4, 2024, https://www.youtube.com/watch?v=IwU0Eqe9v6A.

89. Bernard Marr, "AI In Mental Health: Opportunities and Challenges in Developing Intelligent Digital Therapies," *Forbes*, July 6, 2023, https://www.forbes.com/sites/bernardmarr/2023/07/06/ai-in-mental-health-opportunities-and-challenges-in-developing-intelligent-digital-therapies/?sh=41f2f70e5e10.

90. Ahmed Banafa, "Psychological Impacts of Using AI," OpenMind, May 8, 2023, https://www.bbvaopenmind.com/en/technology/digital-world/psychological-impacts-of-using-ai/.

91. See Jonathan Haidt, "Generation Anxiety: Smartphones Have Created a Gen Z Mental Health Crisis—but There Are Ways to Fix It," *The Guardian*, March 24, 2024, https://www.theguardian.com/books/2024/mar/24/the-anxious-generation-jonathan-haidt-book-extract-instagram-tiktok-smartphones-social-media-screens.

92. Chris Burr, "Concern over Reliance on Digital Technologies for Student Mental Health Care," Alan Turing Institute, November 3, 2022, https://www.turing.ac.uk/news/concern-over-reliance-digital-technologies-student-mental-health-care.

93. Ananya Singh, "Robots, Virtual Intimacy and Sex in Space: Is This the Future of Sex?," Swaddle, January 15, 2023, https://www.theswaddle.com/robots-virtual-intimacy-and-sex-in-space-is-this-the-future-of-sex.

94. Anatoly Kotlyar, "Sextech Startups: Barriers and Perspectives," CX Dojo, accessed May 4, 2024, https://cxdojo.com/sextech-startups.

95. See "Don't Date Robots," YouTube, accessed May 4, 2024, https://www.youtube.com/watch?v=3O3-ngj7I98; "Futurama S03E15—Fry Dates Lucy Liu," YouTube, accessed May 4, 2024, https://www.youtube.com/watch?v=NVs8_DzaRrE.

96. See Jon Haidt and Zach Rausch, "Kids Who Get Smartphones Earlier Become Adults with Worse Mental Health," After Babel, May 15, 2023, https://jonathanhaidt.substack.com/p/sapien-smartphone-report.

97. Singh, "Robots, Virtual Intimacy and Sex in Space." For the record, in 1989, I proposed the thought experiment of a membrane device that could be used sexually by couples over great distances. The purpose of the thought experiment was to consider whether or not such virtual-physical acts should be considered to be acts of infidelity through differing scenarios.

98. Singh, "Robots, Virtual Intimacy and Sex in Space."

99. "Pause Giant AI Experiments: An Open Letter," Future of Life, accessed May 4, 2024, https://futureoflife.org/open-letter/pause-giant-ai-experiments/.

100. Nick Bostrom, "Existential Risk Reduction as Global Priority," *Global Policy* 4, no. 1 (2013): 15, https://onlinelibrary.wiley.com/doi/abs/10.1111/1758-5899.12002.

101. Convergence Analysis (website), accessed May 4, 2024, https://www.convergenceanalysis.org/theory-of-change. Special thanks to Deric Cheng for formulating these passages.

102. Convergence Analysis (website).

103. Nick Bostrom, *Superintelligence* (Oxford: Oxford University Press, 2016), 52.

104. Bostrom, *Superintelligence*, 53.

105. To view more on takeoff speeds, see "An Interactive Model of AI Takeoff Speeds," Epoch AI, accessed May 4, 2024, https://epochai.org/blog/interactive-model-of-takeoff-speeds; Tom Davidson, "What a Compute-Centric Framework Says about Takeoff Speeds," Open Philanthropy, June 27, 2023, https://www.openphilanthropy.org/research/what-a-compute-centric-framework-says-about-takeoff-speeds/.

106. "Recursive Self-Improvement," LessWrong, accessed May 4, 2024, https://www.lesswrong.com/tag/recursive-self-improvement#:~:text=Recursive%20self%2Dimprovement%20refers%20to,functionality%20resulting%20in%20improved%20performance.

107. "Recursive Self-Improvement," LessWrong (Tags), accessed May 4, 2024, https://www.lesswrong.com/tag/recursive-self-improvement/history.

108. DragonGod, "Is 'Recursive Self-Improvement' Relevant in the Deep Learning Paradigm?," LessWrong, April 6, 2023, https://www.lesswrong.com/posts/oyK6fYYnBi5Nx5pfE/is-recursive-self-improvement-relevant-in-the-deep

-learning#:~:text=Eliezer%20Yudkowsky%20argues%20that%20a,in%20the %20next%20step%2C%20and.

109. See Wikipedia, s.v. "Technical Singularity," last modified April 30, 2024, https://en.wikipedia.org/wiki/Technological_singularity.

110. See Marc Andreessen, "The Techno-Optimist Manifesto," Andreessen Horowitz, October 16, 2023, https://a16z.com/the-techno-optimist-manifesto/.

111. Cat Zakrzewski, "Director Christopher Nolan Reckons with AI's 'Oppenheimer Moment,'" *Washington Post*, December 30, 2023, https://www .washingtonpost.com/technology/2023/12/30/nolan-interview-ai-threats/.

112. Deborah Yao, "Going beyond Sci-Fi: Why AI Poses an Existential Threat," AI Business, May 3, 2023, https://aibusiness.com/ml/exactly-how-does -ai-pose-an-existential-threat-.

113. Max Tegmark, "TED Talk: How to Keep AI under Control," YouTube, accessed May 4, 2024, https://www.youtube.com/watch?v=xUNx_PxNHrY.

114. Yao, "Going beyond Sci-Fi."

115. See Ellen Glover, "15 Artificial General Intelligence Companies to Know," Builtin, November 8, 2022, https://builtin.com/artificial-intelligence/ artificial-general-intelligence-companies.

116. Janna Anderson and Lee Rainie, "As AI Spreads, Experts Predict the Best and Worst Changes in Digital Life by 2035," Pew Research Center, June 21, 2023, 40, https://www.pewresearch.org/internet/wp-content/uploads/sites/9/ 2023/06/PI_2023.06.21_Best-Worst-Digital-Life_2035_FINAL.pdf.

117. Pete Syme, "A Future 'God-like AI' Could Destroy Humans or Make Them Obsolete if Not Properly Contained, a Prolific AI Investor Warned," *Business Insider*, April 13, 2023, https://www.businessinsider.com/god-like-ai-needs -safety-net-heated-competition-ai-investor-2023-4.

118. Tima Bansal, "Which Company Will Ensure AI Safety? OpenAI or Anthropic?," *Forbes*, January 16, 2024, https://www.forbes.com/sites/tima bansal/2024/01/16/openai-or-anthropic-which-will-keep-you-more-safe/?sh =3b63df7d5122.

119. Rounak Jain, "OpenAI's New CEO Says Real Reason behind Sam Altman Ouster Not AI Safety: Gives Clue on Future Direction," Nasdaq, November 21, 2023, https://www.nasdaq.com/articles/openais-new-ceo-says -real-reason-behind-sam-altman-ouster-not-ai-safety:-gives-clue-on.

120. Jain, "OpenAI's New CEO Says Real Reason behind Sam Altman Ouster Not AI Safety."

121. Jain, "OpenAI's New CEO Says Real Reason behind Sam Altman Ouster Not AI Safety."

122. See "Developing Beneficial AGI Safely and Responsibly," OpenAI, accessed May 4, 2024, https://openai.com/safety.

123. "Developing Beneficial AGI Safely and Responsibly."

124. Dylan Matthews, "The $1 Billion Gamble to Ensure AI Doesn't Destroy Humanity: The Founders of Anthropic Quit OpenAI to Make a Safe AI

Company. It's Easier Said Than Done," *Vox*, September 25, 2023, https://www .vox.com/future-perfect/23794855/anthropic-ai-openai-claude-2.

125. Eliezer Yudkowsky, "Pausing AI Developments Isn't Enough: We Need to Shut It All Down," *TIME*, March 29, 2023, https://time.com/6266923/ai -eliezer-yudkowsky-open-letter-not-enough/.

126. Yudkowsky, "Pausing AI Developments Isn't Enough."

127. Anthony Aguirre, "Close the Gates to an Inhuman Future: How and Why We Should Choose to Not Develop Superhuman General-Purpose Artificial Intelligence," SSRN, November 17, 2023, https://papers.ssrn.com/sol3/papers .cfm?abstract_id=4608505.

128. Aguirre, "Close the Gates to an Inhuman Future."

129. Justin Shovelain and Elliot Mckernon, "Information-Theoretic Boxing of Superintelligences," LessWrong, November 3, 2023, https://www.less wrong.com/posts/NZP6QvkXryJQFGkLF/information-theoretic-boxing-of -superintelligences-1.

Chapter 4

1. Christopher DiCarlo, *So You Think You Can Think* (Lanham, MD: Rowman & Littlefield, 2020).

2. DiCarlo, *So You Think You Can Think*.

3. DiCarlo, *So You Think You Can Think*.

4. Wikipedia, s.v. "Extraordinary claims require extraordinary evidence," last modified May 1, 2024, https://en.wikipedia.org/wiki/Extraordinary_claims _require_extraordinary_evidence.

5. DiCarlo, *So You Think You Can Think*.

6. Socrates, in Plato's *Republic* (ca. 390 BCE). See also "What Is Morality?," Studocu, accessed May 5, 2024, https://www.studocu.com/en-us/document/ xavier-university/global-business/james-rachels-stuart-rachels-what-is-morality -we-are-discussing-no-small-matter-but-how-we-ought-to-live/17201691.

7. James Rachels, *The Elements of Moral Philosophy*, 3rd ed. (New York: McGraw-Hill, 1999), 19.

8. See Christopher DiCarlo, *How to Become a Really Good Pain in the Ass: A Critical Thinker's Guide to Asking the Right Questions*, 2nd ed. (Amherst, NY: Prometheus, 2021), https://www.amazon.ca/How-Become-Really-Good-Pain/dp/ 163388712X.

9. Soraj Hongladarom, "What Buddhism Can Do for AI Ethics: Buddhism Teaches Us to Focus Our Energy on Eliminating Suffering in the World," *MIT Technology Review*, January 6, 2021, https://www.technologyreview.com/2021/ 01/06/1015779/what-buddhism-can-do-ai-ethics/#:~:text=Thus%2C%20the %20Buddha%20teaches%20that,to%20decrease%20pain%20and%20suffering.

10. Thomas Hobbes, *Leviathan* (New York: Penguin, 1986), 186.

11. Hobbes, *Leviathan*, 188.

12. Hobbes, *Leviathan*, 189.

13. Hobbes, *Leviathan*, 190.

14. Hobbes, *Leviathan*, 190.

15. Hobbes, *Leviathan*, 201.

16. Hobbes, *Leviathan*, 227.

17. Laura D'Olimpio, "Big Thinker: Jeremy Bentham," Ethics Centre, July 25, 2019, https://ethics.org.au/big-thinker-jeremy-bentham/#:~:text= Jeremy%20Bentham%20(1748%E2%80%941832),%2Dbeing%20or%20'utility.

18. See Christopher DiCarlo, "Episode 4: Dr. Peter Singer," *All Thinks Considered* (podcast), February 12, 2024, https://allthinksconsidered.com/2024/02/ 12/episode-4-dr-peter-singer/.

19. Jeremy Bentham, *An Introduction to the Principles of Morals and Legislation* (1789), ed. Jonathan Bennett (n.p.: n.p., 2017), 143–44, https://www.early moderntexts.com/assets/pdfs/bentham1780.pdf.

20. DiCarlo, "Episode 4: Dr. Peter Singer."

21. See *Encylopaedia Britannica Online*, s.v. "Deontological Ethics," last modified March 28, 2024, https://www.britannica.com/topic/deontological-ethics.

22. See Crash Test Dummies, "Superman's Song," YouTube, accessed May 5, 2024, https://www.youtube.com/watch?v=EeyhKWjQaKk.

23. See Stan Lee, "*Amazing Fantasy* (1962) #15," Marvel, accessed May 5, 2024, https://www.marvel.com/comics/issue/16926/amazing_fantasy_1962_15.

24. *Cambridge Dictionary*, s.v. "noblesse oblige," accessed May 5, 2024, https:// dictionary.cambridge.org/dictionary/english/noblesse-oblige.

25. Aristotle, *Nicomachean Ethics* I, Loeb Classical Library, accessed May 5, 2024, https://www.loebclassics.com/view/aristotle-nicomachean_ethics/1926/ pb_LCL073.33.xml?readMode=recto#:~:text=Nicomachean%20Ethics%2C %20I.&text=man%20is%20the%20active%20exercise%20of%20his%20soul's %20faculties%20in,and%20most%20perfect%20among%20them.

26. Rachels, *The Elements of Moral Philosophy*, 176.

27. Rachels, *The Elements of Moral Philosophy*, 176.

28. Rachels, *The Elements of Moral Philosophy*, 178.

29. Forrest E. Baird, *Philosophic Classics: Ancient Philosophy*, vol. 1 (New York: Pearson, 2011), https://books.google.ca/books?id=oe9mDAAAQBAJ&pg =PT542&lpg=PT542&dq=%E2%80%9Cthe+mean+by+reference+to+two+vi ces:+the+one+of+excess+and+the+other+of+deficiency%E2%80%9D&source =bl&ots=8CsyCbD2PX&sig=ACfU3U19P97SU70KqwhfnA3evMdVLPqg0Q &hl=en&sa=X&ved=2ahUKEwjJ75DshLmEAxXg4ckDHUZaBlUQ6AF6BA gmEAM#v=onepage&q=%E2%80%9Cthe%20mean%20by%20reference%20to %20two%20vices%3A%20the%20one%20of%20excess%20and%20the%20other %20of%20deficiency%E2%80%9D&f=false.

30. See "Belmont Report," U.S. Department of Health and Human Services, April 18, 1979, https://www.hhs.gov/ohrp/regulations-and-policy/belmont -report/read-the-belmont-report/index.html.

31. "Universal Declaration of Human Rights," United Nations, accessed May 5, 2024, https://www.un.org/en/about-us/universal-declaration-of-human-rights.

32. "Universal Declaration of Human Rights."

33. See "AI Ethics Guidelines Global Inventory," Algorithm Watch, accessed May 5, 2024, https://inventory.algorithmwatch.org/.

34. "1 Big Thing: UN Official Says Tech Needs to Do No Harm," Axios, February 21, 2024, https://www.axios.com/newsletters/axios-ai-plus-64aa3496 -b559-48bb-af4d-b11abdb91921.html?utm_source=newsletter&utm_medium =email&utm_campaign=newsletter_axioslogin&stream=top.

35. "1 Big Thing."

36. See "The UN Guiding Principles," UN Working Group on Business and Human Rights, accessed May 5, 2024, https://www.ohchr.org/sites/default/files/ Documents/Issues/Business/Intro_Guiding_PrinciplesBusinessHR.pdf.

37. "1 Big Thing."

38. "1 Big Thing."

39. "1 Big Thing."

40. See Taylor Sorsensen et al., "A Road Map to Pluralistic Alignment," arXiv, February 7, 2024, https://arxiv.org/pdf/2402.05070.pdf.

41. IEEE Global Initiative on Ethics of Autonomous and Intelligent Systems, *Ethically Aligned Design: A Vision for Prioritizing Human Well-Being with Autonomous and Intelligent Systems, Version 2* (Piscataway, NJ: IEEE, 2017), 2, https://standards .ieee.org/wp-content/uploads/import/documents/other/ead_v2.pdf.

42. See "The Latest Innovations in Artificial Intelligence: Part 2," iKala, July 21, 2012, https://ikala.ai/2021/07/21/ai-innovations-2/.

43. Isaiah Poritz, "AI-Faked Drake, The Weeknd Song Amps Music Industry's IP Alarm," *Bloomberg Law*, May 2, 2023, https://news.bloomberglaw.com/ip -law/ai-faked-drake-the-weeknd-song-amps-music-industrys-ip-alarm.

44. Poritz, "AI-Faked Drake, The Weeknd Song."

45. See "'I'm Glad I'm Dead' Full Special—Fake George Carlin," YouTube, accessed May 5, 2024, https://www.youtube.com/watch?v=_JgT4Sk6D6c&t =128s.

46. See Christopher DiCarlo, "Episode 3: Kelly Carlin," *All Thinks Considered* (podcast), January 29, 2024, https://allthinksconsidered.com/2024/01/29/epi sode-3-kelly-carlin/.

47. Andrew Dalton, "George Carlin Estate Sues over Fake Comedy Special Purportedly Generated by AI," AP, January 26, 2024, https://apnews.com/ article/george-carlin-artificial-intelligence-special-lawsuit-39d64f728f7a6a621f25 d3f4789acadd.

48. See "*Victor/Victoria*," IMDb, accessed May 5, 2024, https://www.imdb .com/title/tt0084865/.

49. "AI Alignment Proposal #6: Aligning AI Systems to Human Values and Ethics," Medium, July 30, 2023, https://medium.com/@aialignmentproposals/ aligning-ai-systems-to-human-values-and-ethics-5505953a164f.

50. Anthony Aguirre, "Close the Gates to an Inhuman Future: How and Why We Should Choose to Not Develop Superhuman General-Purpose Artificial Intelligence," arXiv, February 1, 2024, 2, https://arxiv.org/pdf/2311.09452.pdf.

51. Aguirre, "Close the Gates to an Inhuman Future."

52. Aguirre, "Close the Gates to an Inhuman Future."

53. For an interesting account on "boxing" superintelligent AI, see Justin Shovelain and Elliot Mckernon, "Information-Theoretic Boxing of Superintelligences," LessWrong, November 30, 2023, https://www.lesswrong.com/posts/NZP6QvkXryJQFGkLF/information-theoretic-boxing-of-superintelligences-1.

54. Max Tegmark and Steve Omohundro, "Provably Safe Systems: The Only Path to Controllable AGI," arXiv, September 5, 2023, https://arxiv.org/abs/2309.01933.

55. This list comes from a talk that Steve Omohundro does online and live. He presented the talk to the team at Convergence Analysis. Steve Omohundro, "Provably Safe Systems: The Only Path to Controllable AGI," YouTube, accessed May 26, 2024, https://www.youtube.com/watch?v=nUrYCUkTFE4.

56. Tegmark and Omohundro, "Provably Safe Systems."

57. Christopher DiCarlo, "Season 2, Episode 2: Dr. Steve Omohundro," *All Thinks Considered* (podcast), May 20, 2024, https://allthinksconsidered.com/.

58. DiCarlo, "Season 2, Episode 2: Dr. Steve Omohundro."

59. DiCarlo, "Season 2, Episode 2: Dr. Steve Omohundro."

60. DiCarlo, "Season 2, Episode 2: Dr. Steve Omohundro."

61. DiCarlo, "Season 2, Episode 2: Dr. Steve Omohundro."

62. DiCarlo, "Season 2, Episode 2: Dr. Steve Omohundro."

63. DiCarlo, "Season 2, Episode 2: Dr. Steve Omohundro."

64. Eliza Strickland, "OpenAI's Moonshot: Solving the AI Alignment Problem: The ChatGPT Maker Imagines Superintelligent AI without Existential Risks," *IEEE Spectrum*, August 31, 2023, https://spectrum.ieee.org/the-alignment-problem-openai.

65. Strickland, "OpenAI's Moonshot."

66. Lyle Moran, "Lawyer Cites Fake Cases Generated by ChatGPT in Legal Brief," Legal Dive, May 30, 2023, https://www.legaldive.com/news/chatgpt-fake-legal-cases-generative-ai-hallucinations/651557/.

67. Moran, "Lawyer Cites Fake Cases."

68. Strickland, "OpenAI's Moonshot."

69. Alan Roose, "Why an Octopus-Like Creature Has Come to Symbolize the State of AI: The Shoggoth, a Character from a Science Fiction Story, Captures the Essential Weirdness of the AI Moment," *New York Times*, June 9, 2023, https://www.nytimes.com/2023/05/30/technology/shoggoth-meme-ai.html.

70. Roose, "Why an Octopus-Like Creature Has Come to Symbolize the State of AI."

71. Roose, "Why an Octopus-Like Creature Has Come to Symbolize the State of AI."

72. Roose, "Why an Octopus-Like Creature Has Come to Symbolize the State of AI."

73. Roose, "Why an Octopus-Like Creature Has Come to Symbolize the State of AI."

74. "How Generative Models Could Go Wrong: A Big Problem Is That They Are Black Boxes," *The Economist*, April 19, 2023, https://www.economist.com/science-and-technology/2023/04/19/how-generative-models-could-go-wrong.

75. Strickland, "'OpenAI's Moonshot."

76. Strickland, "'OpenAI's Moonshot."

77. "How Generative Models Could Go Wrong."

78. "How Generative Models Could Go Wrong."

79. "How Generative Models Could Go Wrong."

80. "How Generative Models Could Go Wrong."

81. See Christopher DiCarlo, "How Problem Solving and Neurotransmission in the Upper Paleolithic Led to the Emergence and Maintenance of Memetic Equilibrium in Contemporary World Religions," in "Bioculture: Evolutionary Cultural Studies," a special evolutionary issue of *Politics and Culture* 1 (April 2010); Christopher DiCarlo, "The Co-evolution of Consciousness and Language and the Development of Memetic Equilibrium," *Journal of Consciousness Exploration & Research* 1, no. 4 (2010); Christopher DiCarlo and John Teehan, "On the Naturalistic Fallacy: A Conceptual Basis for Evolutionary Ethics," *Evolutionary Psychology: An International Journal of Evolutionary Approaches to Psychology and Behavior* 2 (March 2004): 32–46; Christopher DiCarlo, "Abstract: The Influence of Selection Pressures and Secondary Epigenetic Rules on the Cognitive Development of Specific Forms of Reasoning," *Journal of Consciousness Studies: Consciousness Research Abstracts* (2000): 137.

82. See Robin Dunbar, *Grooming, Gossip, and the Evolution of Language* (Cambridge, MA: Harvard University Press, 1998), https://archive.org/details/isbn_9780674363366.

83. DiCarlo, "The Co-evolution of Consciousness."

84. See DiCarlo, "How Problem Solving"; DiCarlo, "The Co-evolution of Consciousness"; DiCarlo and Teehan, "On the Naturalistic Fallacy"; DiCarlo, "Abstract: The Influence of Selection Pressures."

85. Wikipedia, s.v. "Stochastic Parrot," last modified April 17, 2024, https://en.wikipedia.org/wiki/Stochastic_parrot.

86. "Stochastic Parrot."

87. "Stochastic Parrot."

88. DiCarlo, "Episode 4: Dr. Peter Singer."

89. DiCarlo, "Episode 4: Dr. Peter Singer."

90. DiCarlo, "Episode 4: Dr. Peter Singer."

91. Noam Chomsky, "The False Promise of ChatGPT," *New York Times*, March 8, 2023, https://www.nytimes.com/2023/03/08/opinion/noam-chomsky-chatgpt-ai.html.

92. Chomsky, "The False Promise of ChatGPT."

93. Patrick Butlin et al., "Consciousness in Artificial Intelligence: Insights from the Science of Consciousness," arXiv, August 17, 2023, https://doi.org/10.48550/arXiv.2308.08708.

94. Butlin et al., "Consciousness in Artificial Intelligence."

95. Butlin et al., "Consciousness in Artificial Intelligence."

96. Butlin et al., "Consciousness in Artificial Intelligence."

97. Butlin et al., "Consciousness in Artificial Intelligence."

98. Natasha Tiku, "The Google Engineer Who Thinks the Company's AI Has Come to Life," *Washington Post*, June 11, 2022, https://www.washingtonpost.com/technology/2022/06/11/google-ai-lamda-blake-lemoine/.

99. Tiku, "The Google Engineer Who Thinks the Company's AI Has Come to Life."

100. Tiku, "The Google Engineer Who Thinks the Company's AI Has Come to Life."

101. Tiku, "The Google Engineer Who Thinks the Company's AI Has Come to Life."

102. Tiku, "The Google Engineer Who Thinks the Company's AI Has Come to Life."

103. Simon Goldstein and Peter S. Park, "AI Systems Have Learned How to Deceive Humans: What Does That Mean for Our Future?," *The Conversation*, September 4, 2023, https://theconversation.com/ai-systems-have-learned-how-to-deceive-humans-what-does-that-mean-for-our-future-212197.

Chapter 5

1. Carl Sagan, *Pale Blue Dot* (New York: Ballantine, 1997), 200.

2. Michael Veale, Kira Matus, and Robert Gorwa, "AI and Global Governance: Modalities, Rationales, Tensions," *Annual Review of Law and Social Science* 19 (2023): 255–75, https://doi.org/10.1146/annurev-lawsocsci-020223-040749.

3. See appendix A.

4. See "The Statute of the IAEA," International Atomic Energy Agency, accessed May 9, 2024, https://www.iaea.org/about/statute#a1-2.

5. See "Urging an International AI Treaty: An Open Letter," AI Treaty, accessed May 9, 2024, https://aitreaty.org/.

6. Ina Fried, "Exclusive: Public Trust in AI Is Sinking across the Board," Axios, February 4, 2024, https://www.axios.com/2024/03/05/ai-trust-problem-edelman.

7. Isabel Fattal, "How America Stopped Trusting the Experts: A Conversation with Tom Nichols about American Narcissism, the Pandemic, and Declining

Trust," *The Atlantic*, March 22, 2024, https://www.theatlantic.com/newsletters/archive/2024/03/when-experts-fail/677867/.

8. Fattal, "How America Stopped Trusting the Experts."

9. Fattal, "How America Stopped Trusting the Experts."

10. See "Global AI Legislation Tracker," IAPP Research and Insights, accessed May 9, 2024, https://www.skadden.com/-/media/files/publications/2023/12/2024-insights/a-list-of-ai-legislation-introduced-around-the-world.pdf?rev=2349f875f4 6a4acb9ec0639339c4b910&hash=40B5DE79B8DBE1DBCF185E940D67C4AB.

11. See Justin Bullock et al., *The Oxford Handbook of AI Governance* (New York: Oxford University Press, 2024), https://global.oup.com/academic/product/the-oxford-handbook-of-ai-governance-9780197579329?cc=ca&lang=en&.

12. Shana Lynch, "2023 State of AI in 14 Charts," Stanford University, Human-Centered Artificial Intelligence, April 23, 2023, https://hai.stanford.edu/news/2023-state-ai-14-charts.

13. See Benedikt Kohn and Fritz-Ulli Pieper, "AI Regulation around the World," TaylorWessing, May 9, 2023, https://www.taylorwessing.com/en/interface/2023/ai---are-we-getting-the-balance-between-regulation-and-innovation-right/ai-regulation-around-the-world.

14. See "EU AI Act: First Regulation on Artificial Intelligence," European Parliament, last updated December 19, 2023, https://www.europarl.europa.eu/topics/en/article/20230601STO93804/eu-ai-act-first-regulation-on-artificial-intelligence.

15. "EU AI Act."

16. "EU AI Act."

17. "EU AI Act."

18. "EU AI Act."

19. "EU AI Act."

20. "EU AI Act."

21. For those who want to stay current with EU AI Act developments, see *EU Artificial Intelligence Act Newsletter*, accessed May 6, 2024, https://artificialintelligenceact.substack.com/.

22. See "AI Safety Summit Hosted by the UK," AI Safety Summit, accessed May 6, 2024, https://www.aisafetysummit.gov.uk/.

23. Elliot Mckernon, "Update on the UK AI Summit and the UK's Plans," LessWrong, November 10, 2023, https://www.lesswrong.com/posts/gWwMzA gDsskcb2deA/update-on-the-uk-ai-summit-and-the-uk-s-plans. This is an excellent account and synopsis of the main events at the UK AI Summit.

24. "The Bletchley Declaration by Countries Attending the AI Safety Summit, 1–2 November 2023," UK Government, accessed May 6, 2024, https://www.gov.uk/government/publications/ai-safety-summit-2023-the-bletchley-declaration/the-bletchley-declaration-by-countries-attending-the-ai-safety-summit-1-2-november-2023.

25. "The Bletchley Declaration."

26. See "The 17 Goals," United Nations Department of Economic and Social Affairs, accessed May 6, 2024, https://sdgs.un.org/goals.

27. "The Bletchley Declaration."

28. "The Bletchley Declaration."

29. "The Bletchley Declaration."

30. "The Bletchley Declaration."

31. "The Bletchley Declaration."

32. "The Bletchley Declaration."

33. You can see their individual responses here: "Policy Updates," AI Safety Summit, accessed May 6, 2024, https://www.aisafetysummit.gov.uk/policy -updates/#company-policies.

34. I am very grateful to my colleague, Elliot Mckernon, for his review of the UK AI Summit: Elliot Mckernon, "Update on the UK AI Summit and the UK's Plans." The rest of this section refers explicitly to this work.

35. Zach Stein-Perlman, "ARC Evals: Responsible Scaling Policies," Greater Wrong, September 28, 2023, https://www.greaterwrong.com/posts/pnmFBjHt pfpAc6dPT/arc-evals-responsible-scaling-policies.

36. Mckernon, "Update on the UK AI Summit and the UK's Plans."

37. See "Chair's Summary of the AI Safety Summit 2023, Bletchley Park," UK Government, November 2, 2023, https://www.gov.uk/government/publi cations/ai-safety-summit-2023-chairs-statement-2-november/chairs-summary-of -the-ai-safety-summit-2023-bletchley-park.

38. "Chair's Summary of the AI Safety Summit 2023, Bletchley Park."

39. See "Committee on Artificial Intelligence (CAI)," Council of Europe, accessed May 6, 2024, https://www.coe.int/en/web/artificial-intelligence/cai.

40. See "G7 Leaders' Statement on the Hiroshima AI Process," White House Briefing Room, October 30, 2023, https://www.whitehouse.gov/briefing -room/statements-releases/2023/10/30/g7-leaders-statement-on-the-hiroshima -ai-process/.

41. "Call for Partners: Building a Global Challenge on Trust in the Age of Generative AI," OECD, accessed May 6, 2024, https://survey.oecd.org/index .php?r=survey/index&sid=768283&lang=en.

42. "Artificial Intelligence: OECD Principles: How Governments and Other Actors Can Shape a Human-Centric Approach to Trustworthy AI," OECD, accessed May 6, 2024, https://www.oecd.org/digital/artificial-intelligence/ #:~:text=The%20OECD%20Principles%20on%20Artificial,Council%20Recom mendation%20on%20Artificial%20Intelligence.

43. "Ethics of Artificial Intelligence," UNESCO, accessed May 6, 2024, https://www.unesco.org/en/artificial-intelligence/recommendation-ethics.

44. "Chair's Summary of the AI Safety Summit 2023, Bletchley Park."

45. "Office of Science and Technology Policy," White House, accessed May 6, 2024, https://www.whitehouse.gov/ostp/.

46. "OSTP Director: Arati Prabhakar," White House, accessed May 6, 2024, https://www.whitehouse.gov/ostp/directors-office/.

47. "OSTP's Teams: Technology," White House, accessed May 6, 2024, https://www.whitehouse.gov/ostp/ostps-teams/technology/.

48. "Blueprint for an AI Bill of Rights," White House, accessed May 6, 2024, https://www.whitehouse.gov/ostp/ai-bill-of-rights/.

49. "Blueprint for an AI Bill of Rights."

50. "Blueprint for an AI Bill of Rights."

51. See "Executive Order on the Safe, Secure, and Trustworthy Development and Use of Artificial Intelligence," White House Briefing Room, October 30, 2023, https://www.whitehouse.gov/briefing-room/presidential-actions/2023/10/30/executive-order-on-the-safe-secure-and-trustworthy-development-and-use-of-artificial-intelligence/.

52. "Fact Sheet: President Biden Issues Executive Order on Safe, Secure, and Trustworthy Artificial Intelligence," White House Briefing Room, October 30, 2023, https://www.whitehouse.gov/briefing-room/statements-releases/2023/10/30/fact-sheet-president-biden-issues-executive-order-on-safe-secure-and-trustworthy-artificial-intelligence/#:~:text=The%20Executive%20Order%20establishes%20new,around%20the%20world%2C%20and%20more.

53. "Fact Sheet."

54. "Fact Sheet."

55. "Fact Sheet."

56. "Fact Sheet."

57. "Fact Sheet."

58. "Fact Sheet."

59. "Fact Sheet."

60. "Fact Sheet."

61. "Fact Sheet."

62. "Fact Sheet."

63. "Fact Sheet."

64. "Fact Sheet."

65. "Fact Sheet."

66. "Fact Sheet."

67. See "Executive Order on the Safe, Secure, and Trustworthy Development and Use of Artificial Intelligence."

68. Ashyana-Jasmine Kachra, "Making Sense of China's AI Regulations," Holistic AI, February 12, 2024, https://www.holisticai.com/blog/china-ai-regulation#:~:text=In%202022%2C%20China%20passed%20and,technology%20through%20national%2Dlevel%20legislations.

69. "Internet Information Service Algorithmic Recommendation Management Provisions" (initially published August 27, 2021; effective March 1, 2022; English translation of regulations), DigiChina, accessed May 6, 2024, https://

digichina.stanford.edu/work/translation-internet-information-service-algorithmic
-recommendation-management-provisions-effective-march-1-2022/.

70. "Internet Information Service Algorithmic Recommendation Management Provisions."

71. "Internet Information Service Algorithmic Recommendation Management Provisions."

72. "Internet Information Service Algorithmic Recommendation Management Provisions."

73. "Odds and Algorithms: The Intricate Web of Technology in Online Casino Gambling," TMCnet.com, September 6, 2023, https://www.tmcnet.com/topics/articles/2023/09/06/457022-odds-algorithms-intricate-web-technology-online-casino-gambling.htm#:~:text=The%20algorithm%20is%20able%20to,a%20lively%20and%20exciting%20result.

74. Jon Wertheim, "Technology Has Fueled a Sports Betting Boom and a Spike in Problem Gambling, Addiction Therapist Warns," *60 Minutes*, CBS News, February 4, 2024, https://www.cbsnews.com/news/technology-fuels-sports-betting-boom-and-problem-gambling-spike-addiction-therapist-warns-60-minutes-transcript/.

75. Wertheim, "Technology Has Fueled a Sports Betting Boom."

76. Wertheim, "Technology Has Fueled a Sports Betting Boom."

77. "Provisions on the Administration of Deep Synthesis Internet Information Services" (initially published November 25, 2022; effective January 10, 2023), China Law Translate, December 11, 2023, https://www.chinalawtranslate.com/en/deep-synthesis/.

78. Herbert Smith Freehills LLP, "AI-Deep Synthesis Regulations and Legal Challenges: Recent Face Swap Fraud Cases in China," *Digital TMT and Sourcing Notes* (blog), July 27, 2023, https://www.lexology.com/library/detail.aspx?g=1a3455cc-dc4d-4ed0-918a-c3429999c31f#:~:text=China%20July%2027%202023,%2C%20journalism%2C%20or%20political%20satire.

79. "Provisions on the Administration of Deep Synthesis Internet Information Services."

80. "Provisions on the Administration of Deep Synthesis Internet Information Services."

81. "Provisions on the Administration of Deep Synthesis Internet Information Services."

82. "Interim Measures for the Management of Generative Artificial Intelligence" (initially published July 10, 2023), China Law Translate, July 13, 2023, https://www.chinalawtranslate.com/en/generative-ai-interim/.

83. "Interim Measures for the Management of Generative Artificial Intelligence."

84. "Interim Measures for the Management of Generative Artificial Intelligence."

85. See "The Chinese Communist Party's Human Rights Abuses in Xinjang," U.S. Department of State, accessed May 6, 2024, https://2017-2021.state.gov/ccpabuses/#:~:text=The%20U.S.%20Department%20of%20State%2C%20along%20with%20the%20U.S.%20Department,abuses%2C%20including%20forced%20labor%2C%20in.

86. See "China 2023," Amnesty International, accessed May 6, 2024, https://www.amnesty.org/en/location/asia-and-the-pacific/east-asia/china/report-china/.

87. "China 2023."

88. "China 2023."

89. See Center for AI Safety (website), accessed May 6, 2024, https://www.safe.ai/.

90. See Future of Life (website), accessed May 6, 2024, https://futureoflife.org/.

91. See Convergence Analysis (website), accessed May 6, 2024, https://www.convergenceanalysis.org/.

92. Matthijs Maas, "Advanced AI Governance: A Literature Review of Problems, Options, and Proposals," AI Foundations Report 4, SSRN, November 2023, https://papers.ssrn.com/sol3/papers.cfm?abstract_id=4629460.

93. See Markus Anderljung, "Frontier AI Regulation: Managing Emerging Risks to Public Safety," arXiv, last updated November 7, 2023, https://doi.org/10.48550/arXiv.2307.03718.

94. Justin Bullock, interview with author via Zoom, February 23, 2024.

95. Carl Sagan, *Broca's Brain: The Romance of Science* (London: Hodder & Stoughton, 1979).

96. Bullock, interview with author.

97. Bullock, interview with author.

98. Bullock, interview with author.

99. "Specific versus General Principles for Constitutional AI," Anthropic, October 24, 2023, https://www.anthropic.com/news/specific-versus-general-principles-for-constitutional-ai.

100. Stuart Russell, "3 Principles for Creating Safer AI," TED Talk, April 2017, https://www.ted.com/talks/stuart_russell_3_principles_for_creating_safer_ai?language=en.

101. Russell, "3 Principles for Creating Safer AI."

102. Russell, "3 Principles for Creating Safer AI."

103. Russell, "3 Principles for Creating Safer AI."

104. Russell, "3 Principles for Creating Safer AI."

105. Russell, "3 Principles for Creating Safer AI." The theorem is modelled after the Assistance Game. See OECD AI, "Stuart Russell Talks about AI and How to Regulate It at OECD AI Expert Forum," YouTube, accessed May 6, 2024, https://www.youtube.com/watch?v=D5z4p-Ydoew.

106. K. Eric Drexler, "Reframing Superintelligence: Comprehensive AI Services as General Intelligence," Technical Report #2019-1, Future of Humanity Institute, 2019, 27, https://www.fhi.ox.ac.uk/wp-content/uploads/Reframing_Superintelligence_FHI-TR-2019-1.1-1.pdf.

107. Drexler, "Reframing Superintelligence," 33.

108. Christopher DiCarlo, "Episode X: Interview with Dr. Robert Trager," *All Thinks Considered* (podcast), March 27, 2024, https://allthinksconsidered.com/.

109. DiCarlo, "Episode X: Interview with Dr. Robert Trager."

110. DiCarlo, "Episode X: Interview with Dr. Robert Trager."

111. Jack Clark, "Import AI 365," Import AI, March 18, 2024, https://importai.substack.com/p/import-ai-365-wmd-benchmark-amazon?utm_source=post-email-title&publication_id=1317673&post_id=142703743&utm_campaign=email-post-title&isFreemail=true&r=2hvqrb&triedRedirect=true&utm_medium=email.

112. Clark, "Import AI 365."

113. Clark, "Import AI 365."

114. Deric Cheng, "Report: Evaluating an AI Chip Registration Policy," Governance Recommendations Report, Convergence Analysis, April 12, 2024, https://forum.effectivealtruism.org/posts/RuH2anTWnaAZwWz9m/report-evaluating-an-ai-chip-registration-policy.

115. Cheng, "Report: Evaluating an AI Chip Registration Policy."

116. Cheng, "Report: Evaluating an AI Chip Registration Policy."

Chapter 6

1. Zershaaneh Qureshi, "Timelines to Transformative AI: An Investigation," Effective Altruism Forum, March 25, 2024, https://forum.effectivealtruism.org/posts/hzhGL7tb56hG5pRXY/timelines-to-transformative-ai-an-investigation.

2. "Former Google Researcher Timnit Gebru Calls for Stringent AI Regulation," *Economic Times*, May 24, 2023, https://economictimes.indiatimes.com/tech/technology/former-google-researcher-timnit-gebru-calls-for-stringent-ai-regulation/articleshow/100469675.cms.

3. Sawdah Bhaimiya, "A Google Researcher—Who Said She Was Fired after Pointing Out Biases in AI—Says Companies Won't 'Self-Regulate' because of the AI 'Gold Rush,'" *Business Insider*, May 22, 2023, https://www.businessinsider.com/google-former-ai-researcher-on-lack-of-regulation-2023-5.

4. Courtney Rozen, "AI Leaders Are Calling for More Regulation of the Tech: Here's What That May Mean in the US," *Washington Post*, July 27, 2023, https://www.washingtonpost.com/business/2023/07/27/regulate-ai-here-s-what-that-might-mean-in-the-us/f91462c8-2caa-11ee-a948-a5b8a9b62d84_story.html.

5. Ina Fried and Ryan Heath, "1 Big Thing: AI's Mind-Body Problem," Axios, March 15, 2024, https://www.axios.com/newsletters/axios-ai-plus-528b

d5f1–d810-4324-ae4b-00ceff54b121.html?utm_source=newsletter&utm_medium
=email&utm_campaign=newsletter_axioslogin&stream=top.

6. Fried and Heath, "1 Big Thing."

7. Fried and Heath, "1 Big Thing."

8. Ryan Heath, "The Road Map to AI's Next Level Could Be Nature," Axios, March 13, 2024, https://www.axios.com/2024/03/13/verses-ai-artificial -general-intelligence-chatgpt.

9. Heath, "The Road Map to AI's Next Level Could Be Nature."

10. Heath, "The Road Map to AI's Next Level Could Be Nature."

11. Heath, "The Road Map to AI's Next Level Could Be Nature."

12. Tony Ho Tran, "The Radical Movement to Worship AI as a New God: With Algorithms Starting to Govern More and More Aspects of Our Lives, It Was Only a Matter of Time until Someone Started to Deify Them," Daily Beast, February 26, 2023, https://www.thedailybeast.com/the-radical-movement-to -worship-ai-chatbots-like-chatgpt-as-gods.

13. Tran, "The Radical Movement to Worship AI as a New God."

14. Daniel Oberhaus, "Explaining Roko's Basilisk, the Thought Experiment That Brought Elon Musk and Grimes Together," Vice, May 8, 2018, https:// www.vice.com/en/article/evkgvz/what-is-rokos-basilisk-elon-musk-grimes.

15. Oberhaus, "Explaining Roko's Basilisk."

16. Oberhaus, "Explaining Roko's Basilisk."

17. See David Auerbach, "The Most Terrifying Thought Experiment of All Time: Why Are Techno-futurists So Freaked Out by Roko's Basilisk?," Slate, July 17, 2014, https://slate.com/technology/2014/07/rokos-basilisk-the-most -terrifying-thought-experiment-of-all-time.html.

18. Katja Grace et al., "Thousands of AI Authors on the Future of AI," AI Impacts, January 2024, https://aiimpacts.org/wp-content/uploads/2023/04/ Thousands_of_AI_authors_on_the_future_of_AI.pdf.

19. Qureshi, "Timelines to Transformative AI."

20. Qureshi, "Timelines to Transformative AI."

21. Wikipedia, s.v. "Existential Risk from Artificial General Intelligence," accessed May 10, 2024, https://en.wikipedia.org/wiki/Existential_risk_from _artificial_general_intelligence.

22. Corin Katzke, interview with author, March 31, 2024.

23. Maureen Dowd, "Elon Musk's Billion-Dollar Crusade to Stop the AI Apocalypse," Vanity Fair, March 26, 2017, https://www.vanityfair.com/news/ 2017/03/elon-musk-billion-dollar-crusade-to-stop-ai-space-x.

24. See Temple Grandin, "The Grandin Papers," Temple Grandin, Ph.D., accessed May 7, 2024, https://www.templegrandin.com/.

25. David Varga, "Ilya: The AI Scientist Shaping the World," Effective Altru-ism Forum, November 20, 2023, https://forum.effectivealtruism.org/posts/ THfXNTP6YdXnNge8P/ilya-the-ai-scientist-shaping-the-world.

26. Varga, "Ilya."

27. See Robin Hanson and Eliezer Yudkowsky, *The Hanson-Yudkowsky AI-Foom Debate* (San Francisco, CA: Machine Intelligence Research Institute, 2013), https://www.goodreads.com/en/book/show/18489235.

28. Peter High, "Max Tegmark Hopes to Save Us from AI's Worst Case Scenarios," *Forbes*, January 7, 2019, https://www.forbes.com/sites/peterhigh/2019/01/07/max-tegmark-hopes-to-save-us-from-ais-worst-case-scenarios/?sh=2f1bcccc672f.

29. Josh Rosenberg, Ezra Karger, Avital Morris, Molly Hickman, Rose Hadshar, Zachary Jacobs, and Philip Tetlock, "Roots of Disagreement on AI Risk: Exploring the Potential and Pitfalls of Adversarial Collaboration," Forecasting Research Institute, March 11, 2024, https://static1.squarespace.com/static/635693acf15a3e2a14a56a4a/t/65ef1ee52e64b52f145ebb49/1710169832137/AIcollaboration.pdf.

30. Rosenberg et al., "Roots of Disagreement on AI Risk."

31. Rosenberg et al., "Roots of Disagreement on AI Risk."

32. Rosenberg et al., "Roots of Disagreement on AI Risk."

33. Rosenberg et al., "Roots of Disagreement on AI Risk."

34. See Qureshi, "Timelines to Transformative AI."

35. Rosenberg et al., "Roots of Disagreement on AI Risk."

36. See "Our Theory of Change," Convergence Analysis, accessed May 10, 2024, https://www.convergenceanalysis.org/theory-of-change.

37. "Our Theory of Change."

38. "Our Theory of Change."

39. "Our Theory of Change."

40. "Our Theory of Change."

41. "Our Theory of Change."

42. "Our Theory of Change."

43. Octavia Reeve, Anna Colom, and Roshni Modhvadia, "What Do the Public Think about AI? Understanding Public Attitudes and How to Involve the Public in Decision-Making about AI," Ada Lovelace Institute, October 26, 2023, https://www.adalovelaceinstitute.org/evidence-review/what-do-the-public-think-about-ai/.

44. Hillary Clinton, *What Happened* (New York: Simon & Schuster, 2017), 241.

Appendix A

1. Gil Press, "Artificial General Intelligence (AGI) Is a Very Human Hallucination," *Forbes*, March 28, 2023, https://www.forbes.com/sites/gilpress/2023/03/28/artificial-general-intelligence-agi-is-a-very-human-hallucination/?sh=7de2988064f2.

Acknowledgments

The writing of this book has been a wonderful experience and team effort on the part of so many people—from colleagues and academics, to friends and family. First and foremost, I would like to thank the amazing team at Convergence Analysis. It starts with the captain of the team, CEO David Kristoffersson. David has been instrumental in supporting my research and writing. He gave me the time, freedom, and encouragement to produce a work that will hopefully educate and enlighten, but most of all motivate, people to act based on the existential risk of AI. I must also thank Gwyn Glasser, who provided invaluable research for the book. His breadth of knowledge in the field of AI combined with his ability to work at light speed greatly accelerated the completion of the manuscript.

The rest of the team at Convergence is owed considerable gratitude as well. From Dr. Justin Bullock, who provided excellent and engaging insight into the governance of AI, to Dr. Elliot Mckernon, who provided in-depth line editing analysis, to Zershaaneh Qureshi, Corin Katzke, and Deric Cheng, who provided cutting-edge insights from their latest research, I am deeply indebted to your commitment to tackling one of our greatest global threats. And a big shout-out goes to our secretary and treasurer, Harry Day, and our operations director, Mike Keough, who work so diligently behind the scenes to keep the train running smoothly and efficiently. I would also like to thank Justin Shovelain, our chief strategist, and board member Kristian Rönn, for our wonderful monthly meetings and discussions. Thank you both for your support, insights, and encouragement.

I also owe a debt of gratitude to my senior acquisitions editor, Jake Bonar, who had the foresight and ambition to push this project forward, along with my senior production editor, Nicole Carty Myers, my assistant acquisitions editor, Brianna Soubannarath, the copyeditors, and the rest of the staff at Prometheus Books and Rowman & Littlefield. Thank you for your care and attention to making this book better than I ever could have done alone.

And finally, I want to thank my family, starting with my wife Linda, who has been with me for more than thirty years. Thank you for your support. May I one day be able to give you the life you deserve. To my son Jeremy and his wife Jen, and to my son Matt and his partner Jamilee, I have learned the most about life from watching you grow into the amazing men you've become. You have turned out to be the greatest sons a man could wish for. And lastly, I must thank my faithful companion and dog Pyrrho, who gets me out into nature every day so we can clear our heads and embrace the serene pleasure of simply being. You, old friend, and the spirit of Diogenes that dwells within you, have taught me much throughout the years. Thank you.

Index